A Tutorial Introduction to Complex Numbers with MATLAB Applications

Orhan Gazi

Copyright Information

Orhan Gazi

Electrical and Electronics Engineering Department

Ankara Medipol University

Ankara, Turkey

ISBN: 9798340277268

© oG

oG Publications

This work is subject to copyright. All rights are reserved by the Publisher, whether the whole or part of the material is concerned, specifically the rights of translation, reprinting, reuse of illustrations, recitation, broadcasting, reproduction on microfilms or in any other physical way, and transmission or information storage and retrieval, electronic adaptation, computer software, or by similar or dissimilar methodology now known or hereafter developed. The use of general descriptive names, registered names, trademarks, service marks, etc. in this publication does not imply, even in the absence of a specific statement, that such names are exempt from the relevant protective laws and regulations and therefore free for general use. The publisher, the authors, and the editors are safe to assume that the advice and information in this book are believed to be true and accurate at the date of publication. Neither the publisher nor the authors or the editors give a warranty, expressed or implied, with respect to the material contained herein or for any errors or omissions that may have been made. The publisher remains neutral with regard to jurisdictional claims in published maps and institutional affiliations. This imprint is published by Orhan Gazi.

Table of Contents

Preface 5

Chapter-1 6

Introduction to Complex Numbers 6

   1.1 Complex Number 6

   1.2 Arithmetic Operations with Complex Numbers 8

Chapter-2 26

Polar Form of Complex Numbers, Powers and Roots 26

   2.1 Polar Form of Complex Numbers 26

   2.2 Multiplication and Division in Polar Form 42

      2.2.1 Multiplication in Polar Form 42

      2.2.2 Division in Polar Form 44

      2.2.3 de Moivre's formula 46

   2.3 Roots of a Complex Number 51

Chapter-3 59

Introduction to Complex Numbers 59

   3.1 Complex Plane 59

   3.2 Regions in the Complex Plane 61

   3.3 Definitions for Complex Regions 68

   3.4 Line Integral in the Complex Plane 76

      3.4.1 Complex Integral Evaluation 84

Chapter-4 96

Logarithm, Power, Trigonometry of Complex Numbers, Cauchy Theorem 96

   4.1 Logarithm of Complex Numbers 96

   4.2 General Powers 98

   4.3 Trigonometric Complex Functions 101

      4.3.1 Hyperbolic Trigonometric Complex Functions 102

   4.4 Bounds for Complex Integrals 103

   4.5 Cauchy's Integral Theorem 104

      4.5.1 Simple Closed Path 104

      4.5.2 Simply Connected Domain 104

      4.5.3 Cauchy's Integral Theorem 105

      4.5.4 Independence of Path 108

4.6 Cauchy's Integral Theorem for Multiply Connected Domains 109
    4.6.1 Cauchy's Integral Formula 112
    4.6.2 Multiply Connected Domains 118
4.7 Residue Integration 120
    4.7.1 The Residue Theorem 123
4.8 Derivative and Limits of Complex functions 125
    4.8.1 Half-Plane Definitions 126
    4.8.2 Complex Function 126
    4.8.3 Limit, Continuity 127
4.9 Cauchy–Riemann Equations for Complex Numbers 128
Bibliography 130

Preface

Complex numbers is an important subject of mathematics. Complex numbers are very widely used in many engineering fields. Especially in electrical engineering, complex numbers have a wide range of applications. In circuit analysis, signal processing, signals and systems, electromagnetics, analog and digital communication courses complex numbers are used a lot. Without the introduction of complex numbers, it would not be possible to define Fourier transform, Fourier series formulas which are vital subjects of signal processing.

In this book, we explain fundamental concepts of complex numbers. Chapter-1 is devoted to the introduction of complex numbers; arithmetic operations performed on complex numbers are explained in this chapter. Conjugate of complex numbers and properties of conjugate operation is explained in chapter-1. Polar form of complex numbers is explained in chapter-2. Polar form of complex numbers is used especially in signal processing of electrical engineering. Root calculation for complex numbers is performed when designing filters in signal processing. Poles of Butterworth filters are evaluated by finding roots of complex numbers. Complex integration is explained in chapter-3. Complex integration is used in electromagnetic courses of electrical engineering. We cover the other subjects, such as Cauchy's Integral theorem, residue theorem, trigonometric functions of complex numbers, limits and derivatives of complex numbers in chapter-4. Residue theorem is used in many fields of physics and electrical engineering. Residue theorem is used for evaluating the stability of linear control systems in control engineering field of electrical engineering.

Making practice is vital for good learning of a subject. Without making practice, one cannot learn any subject very well. For theoretical topics such as complex numbers, people consider that making practice for such subjects involves only solving problems using pencil and paper. However, any student who just makes practice using pencil and paper may miss some details about the subject. For this reason, in this book, we included MATLAB practices after almost every subject. MATLAB programs are not only useful for learning the subject better, and they are also useful for checking the results of theoretical calculations. Besides, MATLAB programs have graphical outputs which contain information about the computation results and they enhance the learning level of the students.

The book targets university students studying especially electrical or computer engineering. However, this book can be read by anyone interested in MATLAB applications of complex numbers. The book can be adopted as a supplementary book for a semester course on complex numbers or for a course on engineering mathematics.

I dedicate this book to Ankara Medipol University (AMU) which is a place I am happy to be in.

Prof. Dr. Orhan Gazi   September, 2024

Electrical and Electronics Engineering / AMU

Chapter-1

Introduction to Complex Numbers

Abstract: In this chapter we provide fundamental concepts about complex numbers. We start with the definition of complex numbers, and explain the operations performed on complex numbers and explain the modulus, angle, conjugate of complex number. MATLAB applications are provided to comprehend the subjects better.

1.1 Complex Number

Complex numbers are nothing but the elements of the complex field which is derived from real number field. Arithmetic operations are performed using field elements. The field can contain finite number of elements like binary field or it can contain infinite number of elements line real number and complex number fields. Expanding binary fields we can get other fields.

Complex number field is obtained by expanding real number field. And during this expansion, the elements of the complex number field i which is not available in real number field is defined such that
$i^2 = -1$

Complex number field elements, i.e., complex numbers can be defined with the ordered tuple

(x, y)
or they can be defined as

$x + iy$

where x is called the real part of the complex number and y is called imaginary part of the complex number.

The complex numbers are usually denoted by z as in

$z = x + iy \qquad z = (x, y)$

The complex number

$z = x + iy$

can be depicted in complex plane as shown in Figure-1.1 where horizontal axis is used for real parts and vertical axis is used for imaginary parts.

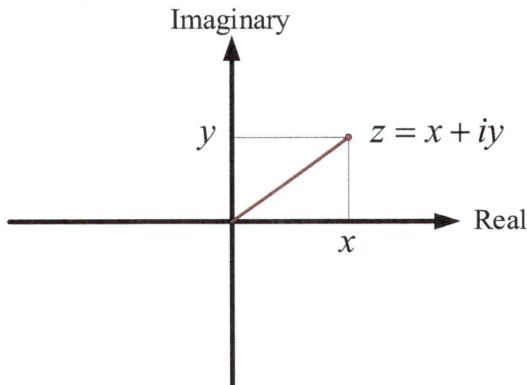

Figure-1.1 Complex plane

For the simplicity of illustration, in our book we will denote the real and imaginary axes of the complex plane by Re and Im as depicted in Figure-1.2.

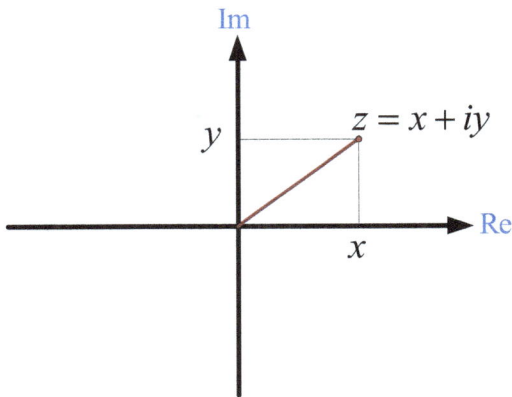

Figure-1.2 Complex plane

The real and imaginary part of the complex number is expressed as

$x = Re\{z\} \quad y = Im\{z\}$

1.2 Arithmetic Operations with Complex Numbers

In this section we explain the arithmetic operations performed with complex numbers.

1.2.1 Addition

Addition of $z_1 = x_1 + iy_1$ and $z_2 = x_2 + iy_2$ is defined as

$$z = z_1 + z_2 = (x_1 + iy_1) + (x_2 + iy_2) \rightarrow z = (x_1 + x_2) + i(y_1 + y_2)$$

or using the paired tuples, we can define addition as

$$z = z_1 + z_2 = (x_1, y_1) + (x_2, y_2) \rightarrow z = (x_1 + x_2, y_1 + y_2)$$

1.2.2 Multiplication

Multiplication of $z_1 = x_1 + iy_1$ and $z_2 = x_2 + iy_2$ is defined as

$$z = z_1 z_2 = (x_1 + iy_1) \times (x_2 + iy_2) \rightarrow$$
$$z = (x_1 x_2 - y_1 y_2) + i(x_1 y_2 + x_2 y_1)$$

Any real number x_1 has zero imaginary part and it can be written as

$$z = x_1 + i0$$

A complex number can have only imaginary part, i.e., we can have

$$z = 0 + iy$$

The real and imaginary part of a complex number can be real numbers as well, for instance

$$z = 2.4 + i(-8.5)$$
which can also be written as
$$z = 2.4 - 8.5i$$

Example-1.1:
$z_1 = 4 + 3i \quad z_2 = 6 - 7i$

$z_1 + z_2 = 4 + 3i + 6 - 7i \rightarrow z_1 + z_2 = 10 - 4i$
$z_1 z_2 = (4 + 3i) * (6 - 7i) \rightarrow$

$z_1 z_2 = 24 + 21 + i(18 - 28) \rightarrow$

$z_1 z_2 = 45 - 10i$

1.2.3 Subtraction

Subtraction of $z_1 = x_1 + iy_1$ and $z_2 = x_2 + iy_2$ is defined as

$$z = z_1 - z_2 = (x_1 + iy_1) - (x_2 + iy_2) \to z = (x_1 - x_2) + i(y_1 - y_2)$$

or using the paired tuples, we can define the subtraction as

$$z = z_1 - z_2 = (x_1, y_1) - (x_2, y_2) \to z = (x_1 - x_2, y_1 - y_2)$$

1.2.4 Division

Division of to complex numbers is performed as

$$z = \frac{x_1 + iy_1}{x_2 + iy_2} = \frac{(x_1 + iy_1)(x_2 - iy_2)}{(x_2 + iy_2)(x_2 - iy_2)} \to$$

$$z = \frac{x_1 x_2 + y_1 y_2}{x_2^2 + y_2^2} + i \frac{x_2 y_1 - x_1 y_2}{x_2^2 + y_2^2}$$

Example-1.2:
$z_1 = 4 + 3i \quad z_2 = 6 - 7i$

$$\frac{z_1}{z_2} = \frac{4 + 3i}{6 - 7i} = \frac{(4 + 3i)(6 + 7i)}{(6 - 7i)(6 + 7i)}$$
where using

$(4 + 3i)(6 + 7i) = 24 + 28i + 18i + 21i^2 \to$

$(4 + 3i)(6 + 7i) = 3 + 46i$
$(6 - 7i)(6 + 7i) = 36 + 42i - 42i + (-49i^2) \to$

$(9 - 2i)(9 + 2i) = 36 + 49$

we get
$$\frac{z_1}{z_2} = \frac{4 + 3i}{6 - 7i} = \frac{3 + 46i}{85} \to \frac{z_1}{z_2} = \frac{3}{85} + \frac{46}{85}i$$

1.3 Defining Complex Numbers in MATLAB

In MCode-1.1 we define complex numbers in MATLAB and get the real and imaginary parts of the complex numbers.

MCode-1.1

```
z1 = 4 + 3i
z2 = complex(4, 3) % second method, z1 = z2

z3 = 6 - 7i
z4 = complex(6, -7)

real(z1) % real part of z1
imag(z1) % imaginary part of z1

real(z2) % real part of z2
imag(z2) % imaginary part of z2
```

1.3.1 Arithmetic Operations with Complex Numbers in MATLAB

In MCode-1, the addition, subtraction, multiplication and division of the complex numbers are done using MATLAB script.

MCode-1.2

```
z1 = 4 + 3i    % i3 gives error
z2 = 6 - 7i    % coefficient is written first

real(z1)       % real part of z1
imag(z1)       % imaginary part of z1

z = z1 + z2    % addition of two complex numbers
z = z1 - z2    % subtraction of two complex numbers

z = z1 * z2    % multiplication of two complex numbers
z = z1 / z2    % division of z1 by z2
```

1.3.2 Complex Number Arrays (Vectors) in MATLAB

In MCode-1.3, we define one complex row vector and one complex column vector.

MCode-1.3

```
v = [3 + 4i   -4 - 3i   1 - 2i   -1 - 1i]
```

```
c = [3 + 4i;   -4 - 3i;   1 - 2i;   -1 - 1i]
```

In MCode-1.4, we define two complex vectors and find their sum, difference, element by element division and element by element multiplication.

MCode-1.4

```
x = [1+2i 3-4i 6+7i 9-8i 10+6i]

y = [1+2i 3-4i 6+7i 9-8i 10+6i]

r1 = x + y    % addition of two complex vectors
r2 = x - y    % subtraction of two complex vectors
r3 = x./y     % division of two complex vectors
r4 = x.*y     % multiplication of two complex vectors
```

1.3.4 Plotting Complex Numbers in MATLAB

In MCode-1.5 we plot complex numbers using MATLAB.

MCode-1.5

```
x = [1+2i 3-4i 6+7i 9-8i 10+6i];

plot(real(x),imag(x),'o','MarkerFaceColor',[0 0 0])

% scatter(real(x),imag(x),'k','filled')

axis equal
grid on
xlabel("Re(x)")
ylabel("Im(x)")
```

When MCode-1.5 is run, we get the graphic in Figure-1.3.

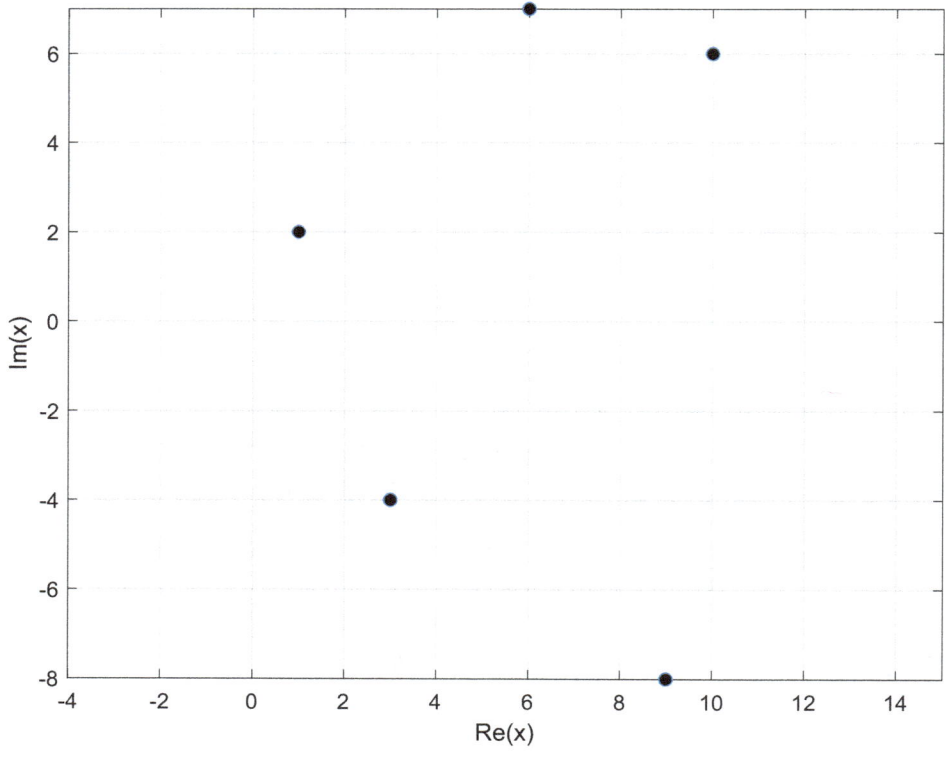

Figure-1.3

1.4 Addition and Subtraction on Complex Plane

In this section, we illustrate the use the addition and subtraction of complex numbers using the complex plane. The addition of two complex numbers is illustrated in Figure-1.3 on a complex plane.

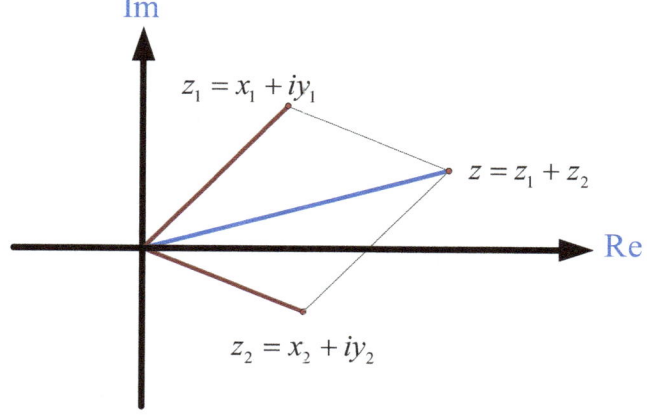

Figure-1.4 Addition of to complex numbers.

The subtraction of two complex numbers is illustrated in Figure-1.5 on a complex plane.

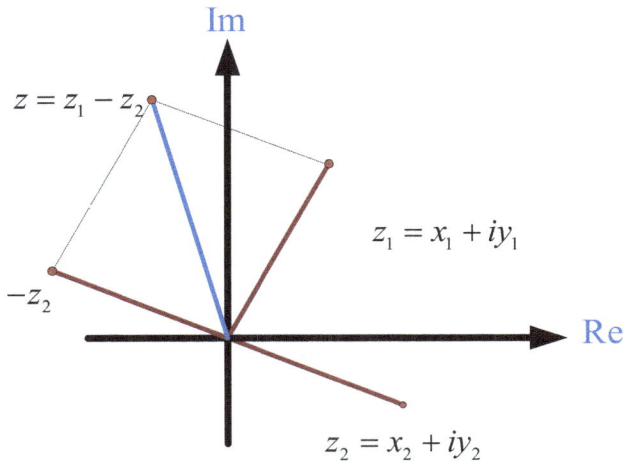

Figure-1.5 Subtraction of two complex numbers.

Complex plane is also called the Argand diagram, after the French mathematician Jean Robert Argand (1768–1822), born in Geneva. His paper on the complex plane appeared in 1806.

In MCode-1.6 and MCode-1.6, addition of two complex number is graphically illustrated in MATLAB.

MCode-1.6

```
z1 = 1+2i;
z2 = 3-4i;
z3 = z1 + z2;

re = [0 real(z1)]
im = [0 imag(z1)]
plot(re, im,'-ko')

legend('z1')
axis equal
grid on
xlabel('Re(x)')
ylabel('Im(x)')
```

When MCode-1.6 is run, we get the graphic in Figure-1.6.

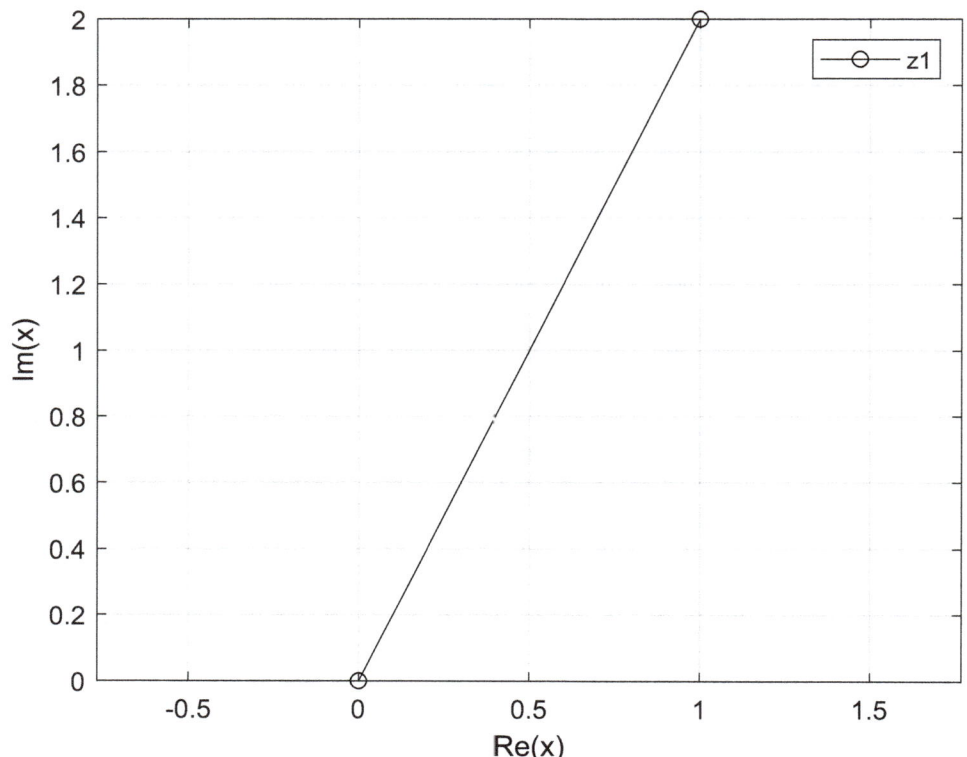

Figure-1.6

MCode-1.7

```
z1 = 1+2i;
z2 = 3-4i;
z3 = z1 + z2;

re = [0 real(z1)]
im = [0 imag(z1)]
plot(re, im,'-ko')
hold on

re = [0 real(z2)]
im = [0 imag(z2)]
plot(re, im,'-b*')

re = [0 real(z3)]
im = [0 imag(z3)]
plot(re, im,'-r<')

legend('z1','z2','z3')
```

```
axis equal
grid on
xlabel('Re(x)')
ylabel('Im(x)')
```

When MCode-1.7 is run, we get the graphic in Figure-1.7.

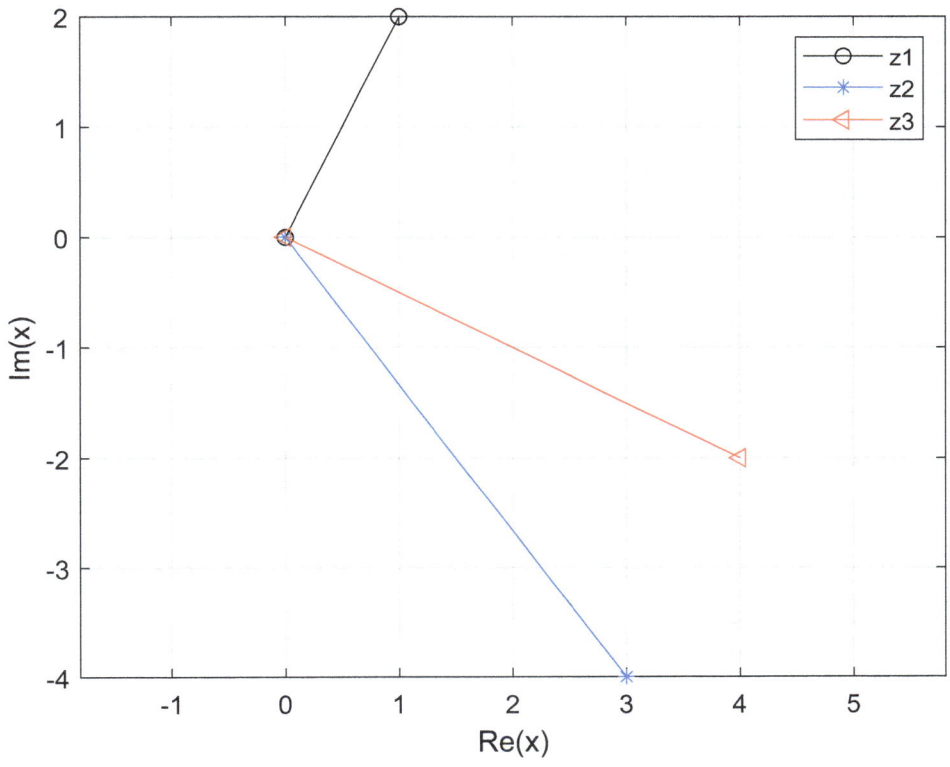

Figure-1.7

1.5 The Modulus or Absolute Value

The absolute value of $z = x + iy$ is defined as

$$|z| = \sqrt{x^2 + y^2}$$

That is,

$$|z|^2 = x^2 + y^2 \rightarrow |z|^2 = (Re\, z)^2 + (Im\, z)^2$$

We have the property

$Re\, z \leq |Re\, z| \leq |z| \quad Im\, z \leq |Im\, z| \leq |z|$

Example-1.3: Let $z = -3 + 2i$

Then,
$|z| = |-3 + 2i| \to |z| = \sqrt{3^2 + 2^2} \to |z| = \sqrt{13}$

Let $z_1 = x_1 + iy_1$ and $z_2 = x_2 + iy_2$, then

$z_1 - z_2 = (x_1 - x_2) + i(y_1 - y_2)$

and

$|z_1 - z_2| = \sqrt{(x_1 - x_2)^2 + (y_1 - y_2)^2}$

which is graphically illustrated in Figure-1.8.

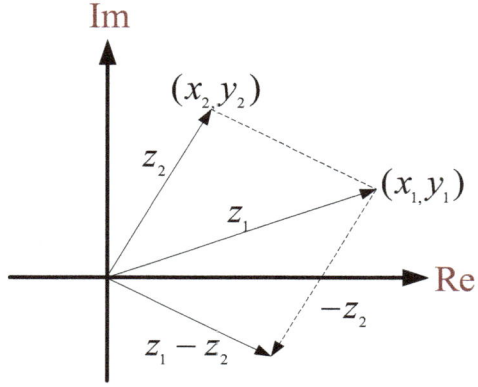

Figure-1.8

In MCode-1.8 and MCode-1.9 we evaluate the absolute value of complex numbers.

MCode-1.8

```
z = -1+3i;

d = abs(z) % absolute value of z, d=3.1623

real(z) % -1

abs(real(z)) % +1
```

```
z1 = -1+3i;

z2 = 4-7i;

d = abs(z1-z2)  %   |z1-z2| --> 11.1803

d
```

Example-1.4: Verify that the equation $|z - 1 + 3i| = 2$ represents the circle whose center is $z_0 = (1, -3)$ and whose radius is $R = 2$.

Solution-1.4: By substituting $z = x + iy$ into

$|z - 1 + 3i| = 2$
we get
$|x + iy - 1 + 3i| = 2$
leading to

$|x - 1 + i(3 + y)| = 2$

which can be written as

$(x - 1)^2 + (y + 3)^2 = 2^2$

which is the equation of a circle with center coordinates $(1, -3)$ and radius 2.

Exercise: Show that $|z| = 3$ represents a circle with radius 3 centered at origin.

Property: Let $z = x + iy$, then we have

$|z| = |-z|$
Proof:
$|-z| = |-x - iy| \rightarrow$

$|-z| = \sqrt{((-x)^2 + (-y)^2)} = \sqrt{x^2 + y^2} = |z|$

Example-1.5: Show that $|z|^2 = |z^2|$

Proof-1.5: Let $z = x + iy$

Then, we have

$$|z| = \sqrt{x^2 + y^2} \rightarrow |z|^2 = x^2 + y^2 \tag{1.1}$$

The expression, z^2 is calculated as

$$z^2 = (x+iy)(x+iy) = x^2 + 2ixy - y^2$$
then,
$$|z^2| = \sqrt{(x^2-y^2)^2 + (2xy)^2} \rightarrow$$

$$|z^2| = \sqrt{x^4 + y^4 - 2x^2y^2 + 4x^2y^2} \rightarrow$$

$$|z^2| = \sqrt{x^4 + 2x^2y^2 + y^4}$$

$$|z^2| = \sqrt{(x^2+y^2)^2}$$

$$|z^2| = x^2 + y^2 \tag{1.2}$$

When (1.1) is compared to (1.2), we see that they are the same, Hence,

$$|z|^2 = |z^2|$$

In MCode-1.10 we verify the property
$|z|^2 = |z^2|$
using MATLAB programming.

MCode-1.10

```
% Show that |z|^2=|z^2 |

z = 5-9i;

d1 = (abs(z))^2  % d1 = 106

d2 = abs(z*z)    % d2 = 106
```

1.6 Complex Conjugate Numbers

The complex conjugate of a complex number $z = x + iy$ is defined by

$$\bar{z} = x - iy$$

Example-1.6: If $z = 5 + 2i$ then $\bar{z} = 5 - 2i$ which is graphically illustrated in Figure-1.9.

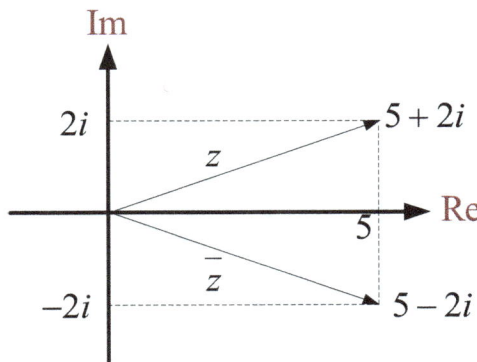

Figure-1.9

Example-1.7: Find $z\bar{z}$, if

$$z = x + iy \quad \bar{z} = x - iy$$

Solution:
$z\bar{z} = (x + iy)(x - iy) \rightarrow$
$z\bar{z} = x^2 - ixy + iyx - i^2y^2 \rightarrow$
$z\bar{z} = x^2 + y^2$

In MCode-1.11, MCode-1.12 use conjugate function of MATLAB.

MCode-1.11

```
z1 = 5-9i

z2 = conj(z1)  % z2 = 5 + 9i
```

MCode-1.12

```
z = 5-9i

d = z * conj(z)   % d = 106
```

1.6.1 Properties of Complex Conjugate

Some properties of complex conjugate can be written as

$$(z_1 + z_2) = \bar{z}_1 + \bar{z}_2 \qquad (z_1 - z_2) = \bar{z}_1 - \bar{z}_2$$

$$\overline{(z_1 z_2)} = \bar{z}_1 \bar{z}_2 \qquad \overline{\left(\frac{z_1}{z_2}\right)} = \frac{\bar{z}_1}{\bar{z}_2}$$

$$Re\, z = x = \frac{1}{2}(z + \bar{z}) \qquad Im\, z = y = \frac{1}{2i}(z - \bar{z})$$

$$z\bar{z} = |z|^2$$

In MCode-1.13, 14, 15 we illustrate the properties of complex conjugate using MATLAB programming.

MCode-1.13

```
z1 = 5-9i;
z2 = 3+7i;

r1 = conj(z1+z2);
r2 = conj(z1)+conj(z2);

r1-r2   % result is 0, since r1 = r2

r3 = conj(z1-z2);
r4 = conj(z1)-conj(z2);

r3-r4   % result is 0, since r3 = r4
```

MCode-1.14

```
z1 = 5-9i;
z2 = 3+7i;

r1 = conj(z1*z2);
r2 = conj(z1)*conj(z2);

r1-r2   % result is 0, since r1 = r2
```

```
r3 = conj(z1/z2);
r4 = conj(z1)/conj(z2);

r3-r4   % result is 0, since r3 = r4
```

MCode-1.15

```
z = 5-9i;
re = real(z);
im = imag(z)

r1 = 0.5*(z+conj(z));
i1 = -0.5i*(z-conj(z));

re-r1 % result is 0 since re = r1
im-i1 % result is 0 since im = i1
```

Example-1.8: Let $z_1 = 4 + 3i$ and $z_2 = 2 + 5i$

Then,

$$Re\ z_1 = \frac{1}{2}(z_1 + \bar{z}_1) \rightarrow Re\ z_1 = \frac{1}{2}(4 + 3i + 4 - 3i) = 4$$

$$Im\ z_1 = \frac{1}{2i}(z_1 - \bar{z}_1) \rightarrow$$

$$Im\ z_1 = \frac{1}{2i}(4 + 3i - (4 - 3i)) \rightarrow$$

$$Im\ z_1 = \frac{1}{2i}(6i) \rightarrow Im\ z_1 = 3$$

Example-1.9: Let $z_1 = 4 + 3i$ and $z_2 = 2 + 5i$

Then,

$$\overline{z_1 z_2} = \overline{(4 + 3i)(2 + 5i)} = \overline{(-7 + 26i)} = -7 - 26i$$

$$\bar{z}_1 \bar{z}_2 = (4 - 3i)(2 - 5i) \rightarrow -7 - 26i$$

Exercise:

1) Let $z_1 = -2 + 11i \quad z_2 = 2 + i$

Show the results in $x + iy$ form

Find

a) $z_1 z_2, \; \bar{z_1} z_2$ b) $Re\{z_1^2\}, \; (Re\{z_1\})^2$

c) $1/z_1$ d) $Im\left(\dfrac{1}{\bar{z}^2}\right)$

1.7 Triangle Inequality

For two complex numbers z_1 and z_2, we have

$$|z_1 + z_2| \leq |z_1| + |z_2|$$

Example-1.10: Show that

$$|z_1 + z_2| \geq ||z_1| - |z_2||$$

Proof-1.10:
$$|z_1| = |(z_1 + z_2) + (-z_2)|$$

where employing the triangle inequality, we get

$$|z_1| = |(z_1 + z_2) + (-z_2)| \leq |z_1 + z_2| + |-z_2|$$

from which we obtain,
$$|z_1| - |z_2| \leq |z_1 + z_2|$$

i.e.,

$$|z_1 + z_2| \geq |z_1| - |z_2| \tag{1.3}$$

Now we start with z_2, i.e.,

$$|z_2| = |(z_1 + z_2) + (-z_1)|$$

where employing the triangle inequality, we get

$$|z_2| = |(z_1 + z_2) + (-z_1)| \leq |z_1 + z_2| + |-z_1|$$

from which we obtain
$$|z_2| - |z_1| \leq |z_1 + z_2|$$

i.e.,
$$|z_1 + z_2| \geq |z_2| - |z_1| \tag{1.4}$$
Thus, we obtained two results

$$|z_1 + z_2| \geq |z_1| - |z_2|$$

$$|z_1 + z_2| \geq |z_2| - |z_1|$$

We can combine these two results under a single expression as

$$|z_1 + z_2| \geq ||z_1| - |z_2||$$

where the expression
$$||z_1| - |z_2||$$

equals $|z_1| - |z_2|$ when $|z_1| > |z_2|$

and it equals $-(|z_1| - |z_2|) = |z_2| - |z_1|$ when $|z_1| < |z_2|$

Note: $\quad |a| = a$ if $a > b$
and
$\quad\quad |a| = -a$ if $a < b$

In MCode-1.16 and MCode-1.17 triangle inequality is verified using MATLAB programming.

MCode-1.16

```
% triangle inequality
% |z_1 + z_2 | ≤ |z_1 | + |z_2 |

z1 = 5-9i;
z2 = 4+7i;

z = z1+z2;
```

```
disp('|z1+z2|')
abs(z1+z2)        % = 9.2195

disp('|z1|+|z2|')
abs(z1)+abs(z2)   % = 18.3579
```

MCode-1.17

```
% |z_1+z_2 |≥|(|z_1 |-|z_2 |)|

z1 = 5-9i;
z2 = 4+7i;

z = z1+z2;
disp('|z1+z2|')
abs(z1+z2)        % = 9.2195

disp('|(|z1|-|z2|)|')
abs(abs(z1)-abs(z2))   % = 2.23
```

Generalized Triangle Inequality

Triangle inequality can be generalized as

$$|z_1 + z_2 + ... z_n| \leq |z_1| + |z_2| + ... + |z_n|$$

Example-1.11: Let z be a complex number such that $|z| = 2$. Then for the expression

$$|3 + z + z^2|$$

we can write the inequality

$$|3 + z + z^2| \leq 3 + |z| + |z^2|$$

where $|z^2|$ can be calculated using
$$|z^2| = |z|^2$$
and $|z| = 2$ as

$$|z^2| = 4$$
Then,

$|3 + z + z^2| \leq 3 + |z| + |z^2|$
can be written as

$|3 + z + z^2| \leq 3 + 2 + 4 \rightarrow |3 + z + z^2| \leq 9$

In MCode-1.18 generalized triangle inequality is verified using MATLAB programming.

MCode-1.18

```matlab
% |z_1+z_2+z_3|≤|z_1|+|z_2|+|z_2|

% |3+z+z^2|  ≤   3+|z|+|z^2|

z = 5-9i;

disp('|3+z+z^2|');
abs(3+z+z*z)    % =   110.0227

disp('3+|z|+|z^2|');
3+abs(z)+abs(z*z)    % = 119.2956
```

Chapter-2

Polar Form of Complex Numbers, Powers and Roots

Abstract: In this chapter we explain the polar form of complex numbers. Polar forms of complex numbers are used when roots and powers of complex numbers are calculated. Roots of complex numbers are calculated in digital signal processing while designing analog filters such as Butterworth filter.

2.1 Polar Form of Complex Numbers

The complex number
$z = x + iy$

is shown in the Figure-2.1.

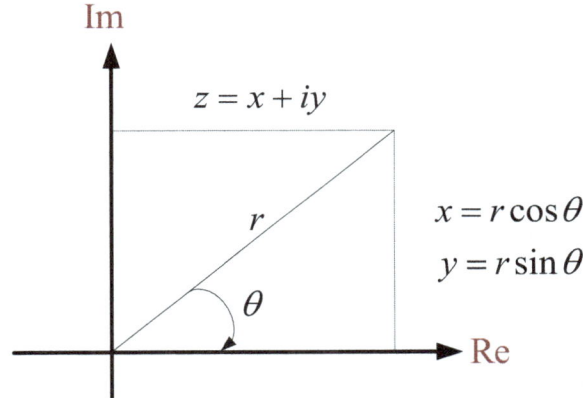

Figure-2.1

From triangle it is seen that

$x = r\cos\theta \quad y = r\sin\theta$
where
$r = \sqrt{x^2 + y^2}$
which can be written as

$r = |z| = \sqrt{z\bar{z}}$

Then, we have
$$z = x + iy \to$$
$$z = r\cos\theta + ir\sin\theta \to$$
$$z = r(\cos\theta + i\sin\theta)$$
The expression
$$z = r\cos\theta + ir\sin\theta$$

can also be written as
$$z = re^{i\theta}$$

where
$$e^{i\theta} = \cos\theta + i\sin\theta$$

The expression
$$z = r(\cos\theta + i\sin\theta)$$
or the one
$$z = re^{i\theta}$$

is called the polar form of a complex number.

The angle θ is called the argument of the complex number and it is calculated as

$$\tan\theta = \frac{y}{x} \to \arg z = \theta = \arctan\frac{y}{x}$$

Since tangent is periodic with periodic 2π, argument of the complex number, there are infinitely many θ values. The value of θ between $-\pi$ and π is called the principal argument of the complex number and it is defined as

$$-\pi \le \theta = Arg\,z < \theta$$

The argument $\arg z$ can be written in terms of principal argument as

$$\arg z = Arg\,z + k2\pi \quad k \in Z$$

where k is an integer.

Example-2.1: Find the polar form of the complex number

$$z = 1 + i$$

and determine its principal argument and general argument.

Solution-2.1: The complex number is shown in Figure below

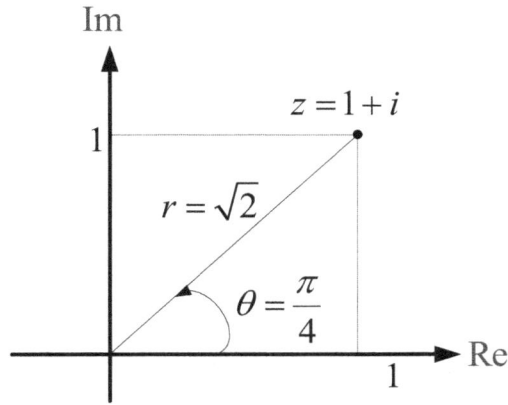

Figure-2.2

Using Figure we can write the polar form of complex number

$z = 1 + i$
using
$z = |z|(\cos \theta + i\sin \theta)$
as
$z = \sqrt{2}\left(\cos\dfrac{\pi}{4} + i\sin\dfrac{\pi}{4}\right)$

which can also be written as

$z = \sqrt{2}\, e^{i\frac{\pi}{4}}$

The principal argument is
$Arg\ z = \dfrac{\pi}{4}$

and the general argument is

$\arg z = Arg\ z + k2\pi$
where k is an integer.

Note that the principal argument $Arg\ z$ is between $-\pi$ and π

Polar Plot in MATLAB

The graphic for the polar form of a complex number can be plotted using **polarplot** function of the MATLAB. Polarplot function most commonly can be used as

$$\text{polarplot(Z)} \qquad \text{polarplot(TH,R)}$$

where Z is the complex number and TH, R are the angle in radian and absolute value of the complex number. When polarplot(Z) is used, TH and R calculated as

$$\text{TH=angle(Z)} \quad \text{R=abs(Z)}$$

In MCode-2.1, 2.2, 2.3 we provide MATLAB examples about the use of polarplot.

MCode-2.1

```
% z = r * e^jθ

z = complex(1,1);  % z = 1+j

abs_z = abs(z)     % 1.4141

theta = angle(z)   % pi / 4 = 0.7854

theta_degree = angle(z)*180/pi  % 45 degree


z1 = abs_z*exp(j*theta)   % z1 = z

z-z1 % 0
```

MCode-2.2

```
% z = r * e^jθ

z = complex(1,1);  % z = 1+j

abs_z = abs(z);    % 1.4141

theta = angle(z);  % pi / 4 = 0.7854

polarplot(theta,abs_z,'r *')
```

When MCode-2.2 is run, we get the graphic in Figure-2.3.

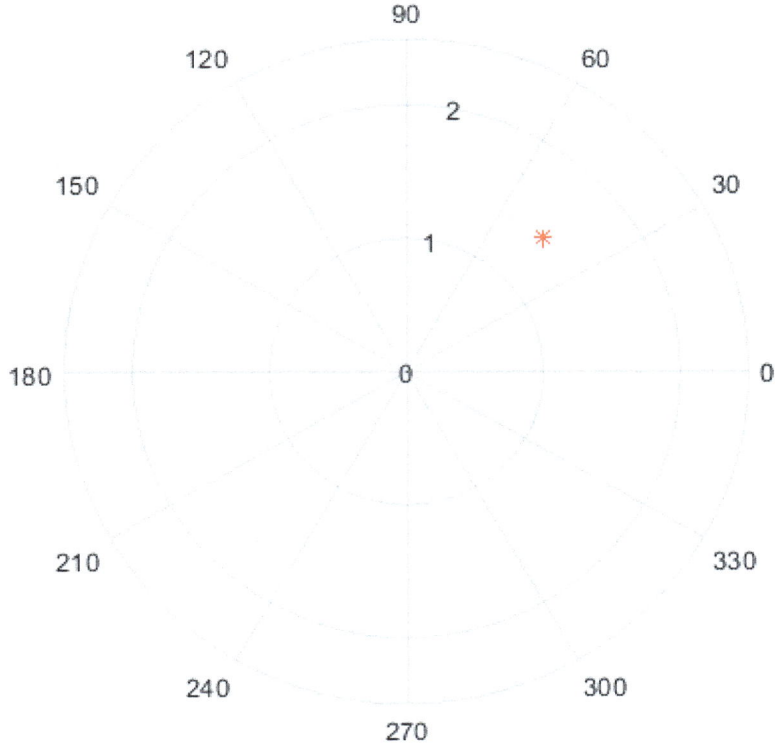

Figure-2.3

MCode-2.3

```
clc; clear all;

theta = linspace(0, 2*pi, 10);
rho = linspace(0, 20, 10);

polarplot(theta, rho, '-o')
```

When MCode-2.3 is run, we get the graphic in Figure-2.4.

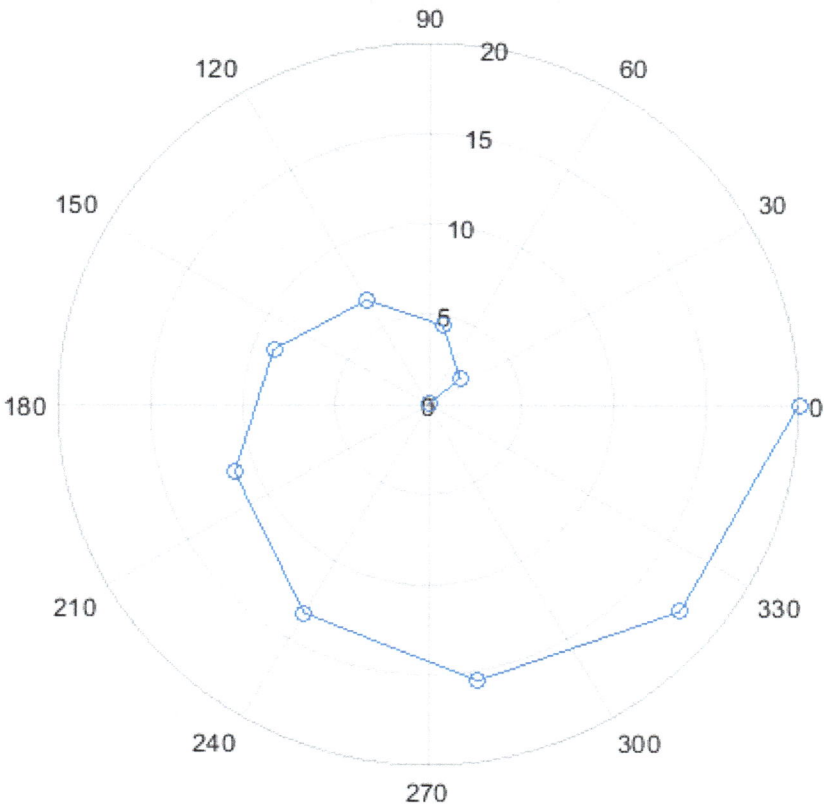

Figure-2.4

Example-2.2: Find the polar form of the complex number

$z = -1 - i$

and determine its principal argument and general argument.

Solution-2.2: The complex number is shown in Figure-2.5.

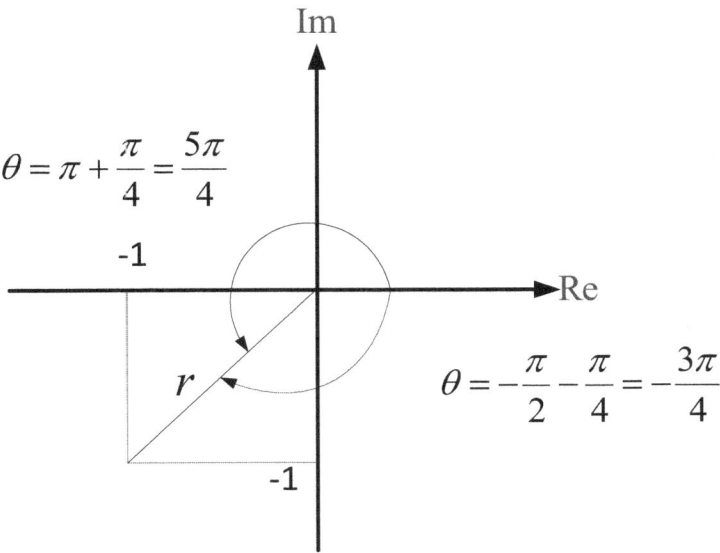

Figure-2.5

The counter clockwise angle is $5\pi/4$ which is outside the interval $[-\pi...\pi)$

Then, we consider clockwise angle $-3\pi/4$ which falls into $[-\pi...\pi)$

Thus, the principal argument is
$$Arg\ z = -\frac{3\pi}{4}$$

and the general argument is
$$\arg z = Arg\ z + k2\pi \rightarrow \arg z = -\frac{3\pi}{4} + k2\pi$$

The polar form of the complex number is

$$z = \sqrt{2}\left(\cos\left(-\frac{3\pi}{4}\right) + i\sin\left(-\frac{3\pi}{4}\right)\right)$$

which can be written as
$$z = \sqrt{2}e^{-i\frac{3\pi}{4}}$$
which can be stated as
$$z = \sqrt{2}\exp\left(-i\frac{3\pi}{4}\right)$$

The polar form can be written in its most general form as

$$z = \sqrt{2}\left(\cos\left(-\frac{3\pi}{4} + k2\pi\right) + i\sin\left(-\frac{3\pi}{4} + k2\pi\right)\right)$$

In MCode-2.4 we plot the polar form of $z = -1 - i$ using MATLAB.

MCode-2.4

```
% z = r * e^jθ
z = complex(-1,-1)  % z = -1-j
abs_z = abs(z)      % 1.4141
theta = angle(z)    % -3*pi / 4 = -2.3562
polarplot(theta,abs_z,'r *')
```

When MCode-2.4 is run, we get the graphic in Figure-2.6.

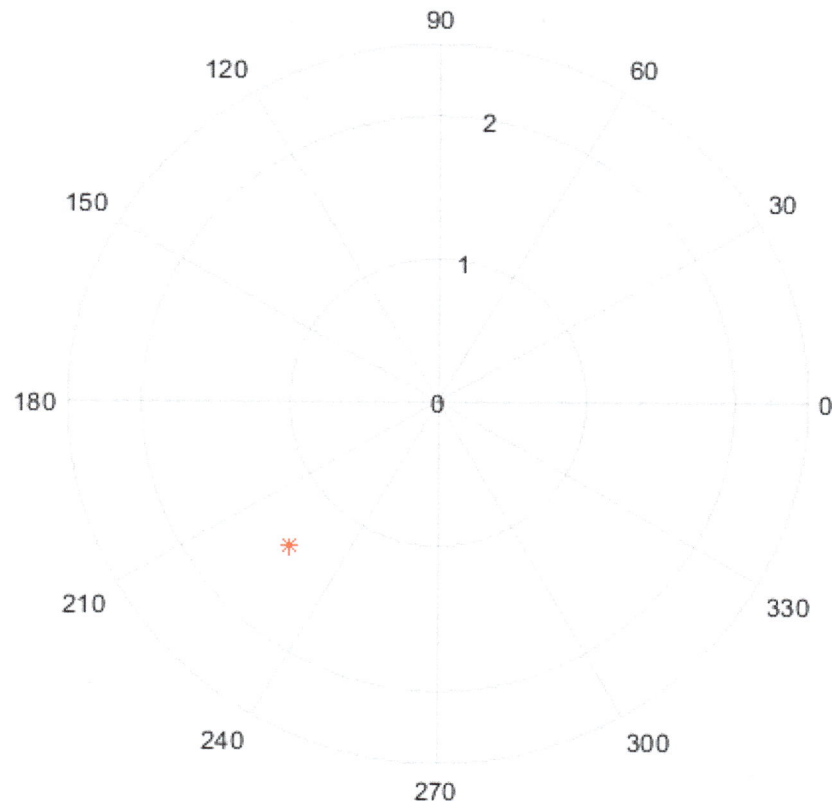

Figure-2.6

Example-2.3: Obtain the polar form of

$z = 1 - i$

Find the principal and the general argument of the complex number.

In MCode-2.4 we plot the polar form of $z = 1 - i$ using MATLAB.

MCode-2.5

```
% z = r * e^jθ
z = complex(1,-1)   % z = 1-j
abs_z = abs(z)      % 1.4141
theta = angle(z)    % -pi / 4 = -0.7854
polarplot(theta,abs_z,'r *')
```

When MCode-2.5 is run, we get the graphic in Figure-2.7.

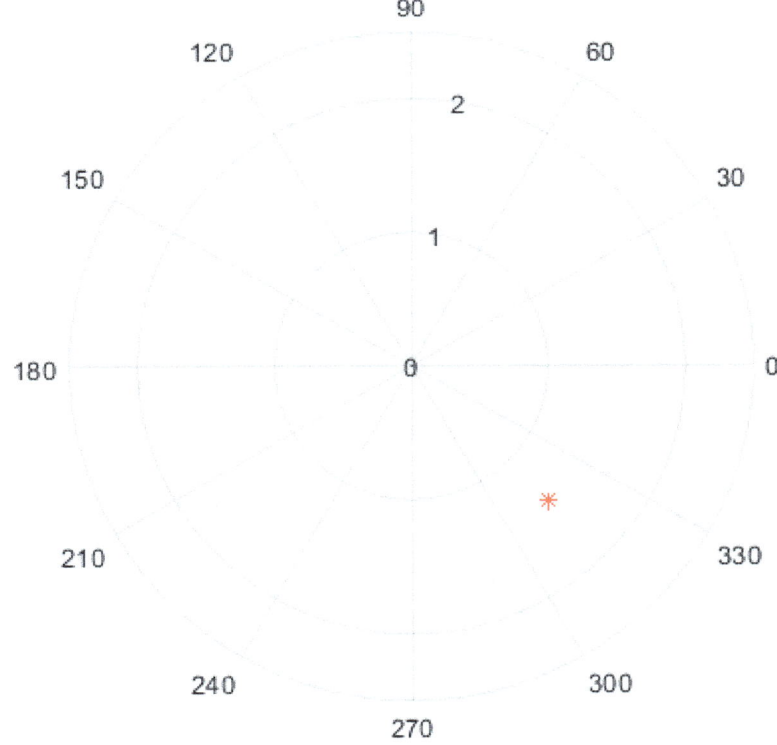

Figure-2.7

In MCode-2.6, we plot a number of complex number in polar coordinates.

MCode-2.6

```
theta = 0 : pi/32 : pi/4;
r = sin(theta);

polarplot(theta, r, 'r -x');
```

When MCode-2.6 is run, we get the graphic in Figure-2.8.

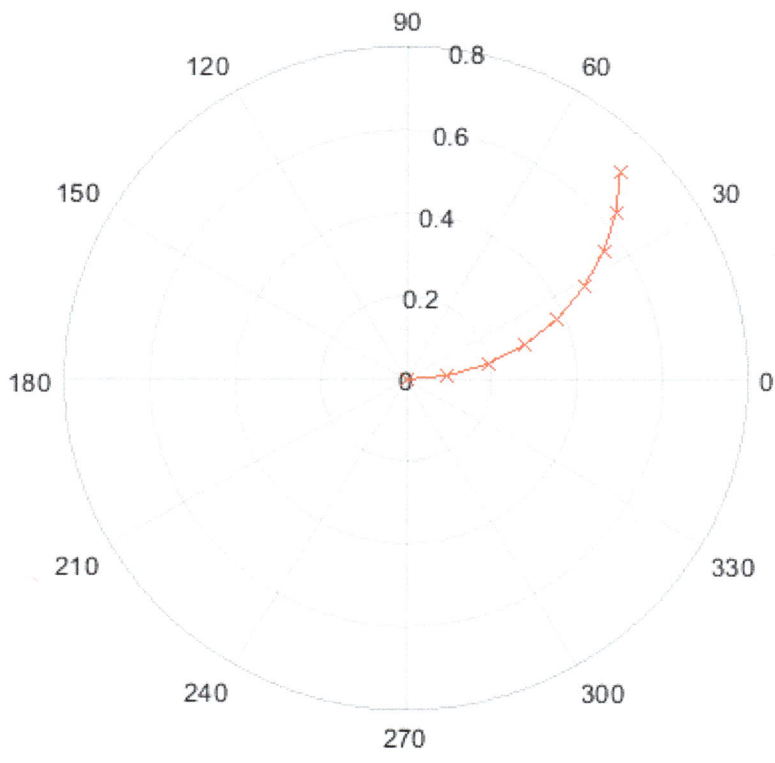

Figure-2.8

In MCode-2.7, the previous example is continued with more complex numbers.

MCode-2.7

```
theta = 0 : pi/32 : pi/4;
r = sin(theta);

polarplot(theta, r, 'r -x');

hold on;

theta = pi/4 : pi/32 : pi/2;
r = sin(theta);

polarplot(theta, r, 'b -*');
```

When MCode-2.7 is run, we get the graphic in Figure-2.9.

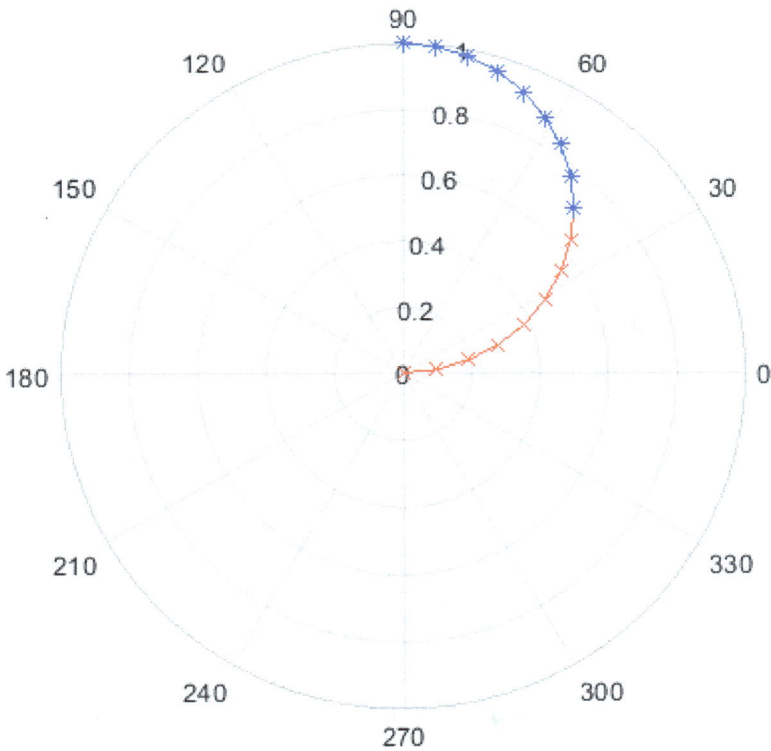

Figure-2.9

In MCode-2.8, 2.9 we illustrate the use of the polarplot function.

MCode-2.8

```
theta = 0 : pi/32 : pi/4;
r = sin(2*theta);

polarplot(theta, r, 'r -x');
```

When MCode-2.8 is run, we get the graphic in Figure-2.10.

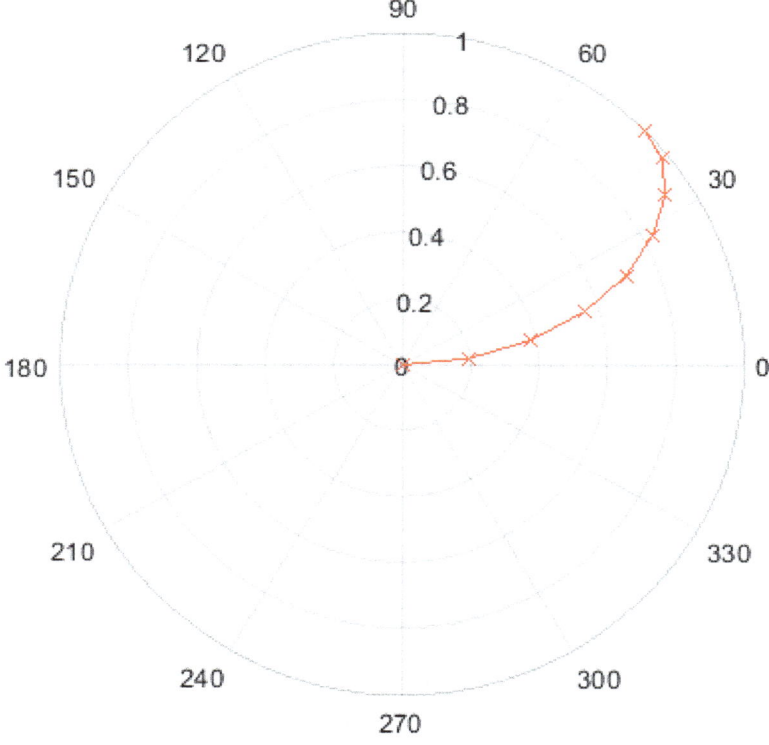

Figure-2.10

MCode-2.9

```
theta = 0 : pi/32 : pi/4;
r = sin(2*theta);

polarplot(theta, r, 'r -x');

hold on;

theta = pi/4 : pi/32 : pi/2;
r = sin(2*theta);

polarplot(theta, r, 'b -*');
```

When MCode-2.9 is run, we get the graphic in Figure-2.11.

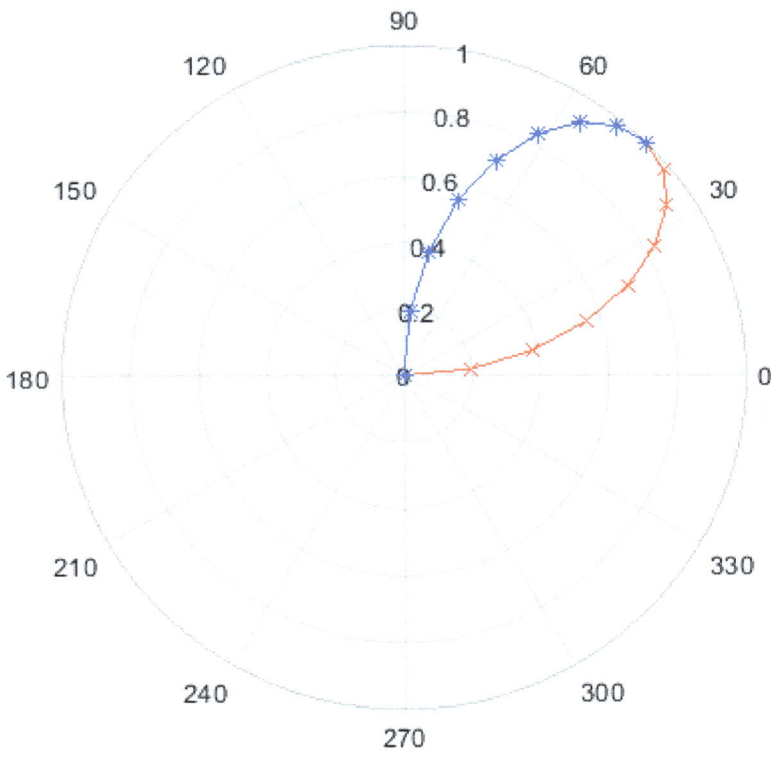

Figure-2.11

For
$0 \leq \theta < 2\pi$

The tip of the magnitude of the complex number

$z = re^{i\theta}$

lies on a circle as shown in the Figure below

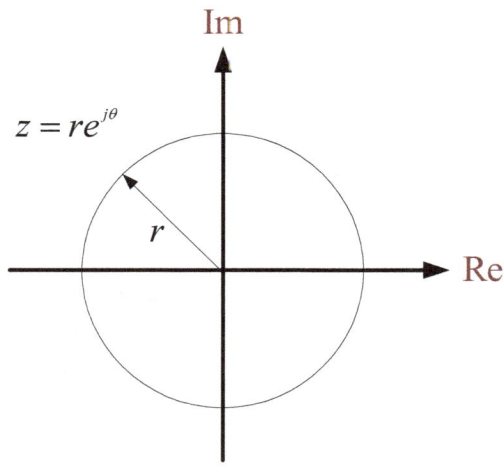

Figure-2.12

In MCode-2.10, we draw the graph of $z = re^{i\theta}$ using MATLAB.

```
MCode-2.10
```

```
r = 2.5;

theta = 0: pi/10:2*pi;

z = r * exp(j*theta);

abs_z = abs(z)    % 1.4141

theta = angle(z)  % -pi / 4 = -0.7854

polarplot(theta,abs_z,'r -x')
```

When MCode-2.10 is run, we get the graphic in Figure-2.13.

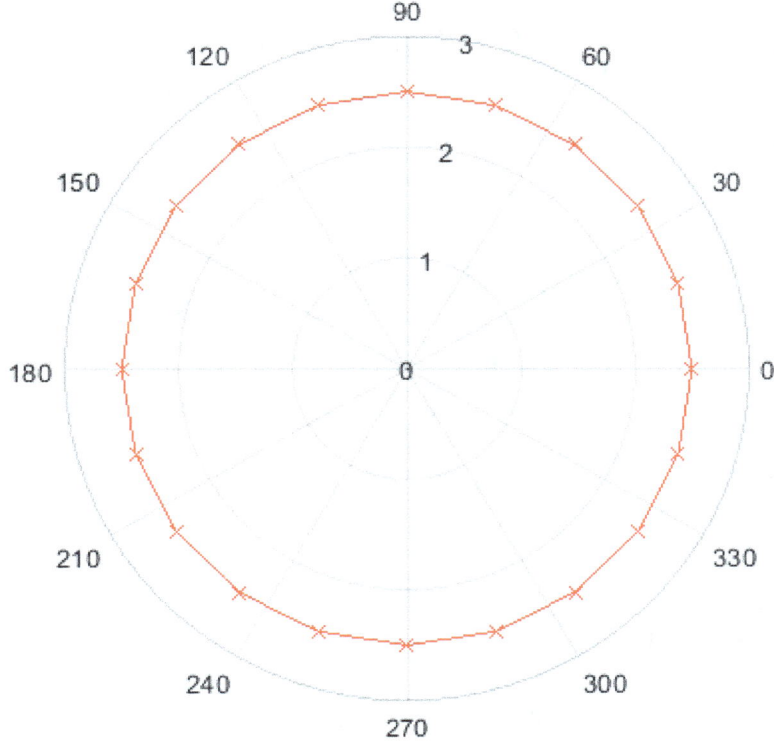

Figure-2.13

The graph
$$z = z_0 + re^{i\theta}$$

is similar to the above Figure, but the center of the circle is z_0 in this case

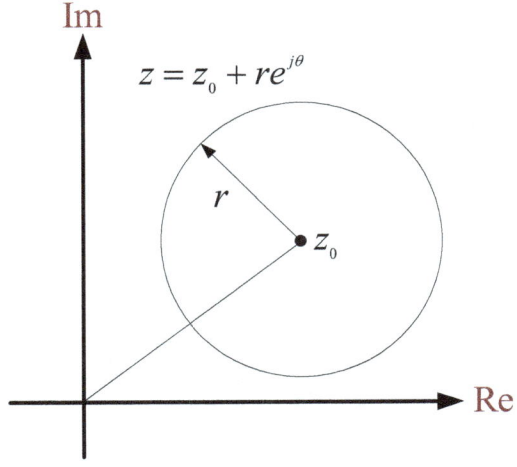

Figure-2.14

In MCode-2.10, we draw the graph of $z = z_0 + re^{i\theta}$ using MATLAB.

MCode-2.11

```
r = 2.5;
z0 = 3 + 4j;
theta = 0: pi/10:2*pi;
z = z0 + r * exp(j*theta);
abs_z = abs(z)   % 1.4141
theta = angle(z) % -pi / 4 = -0.7854
polarplot(theta,abs_z,'r -x')
```

When MCode-2.11 is run, we get the graphic in Figure-2.15.

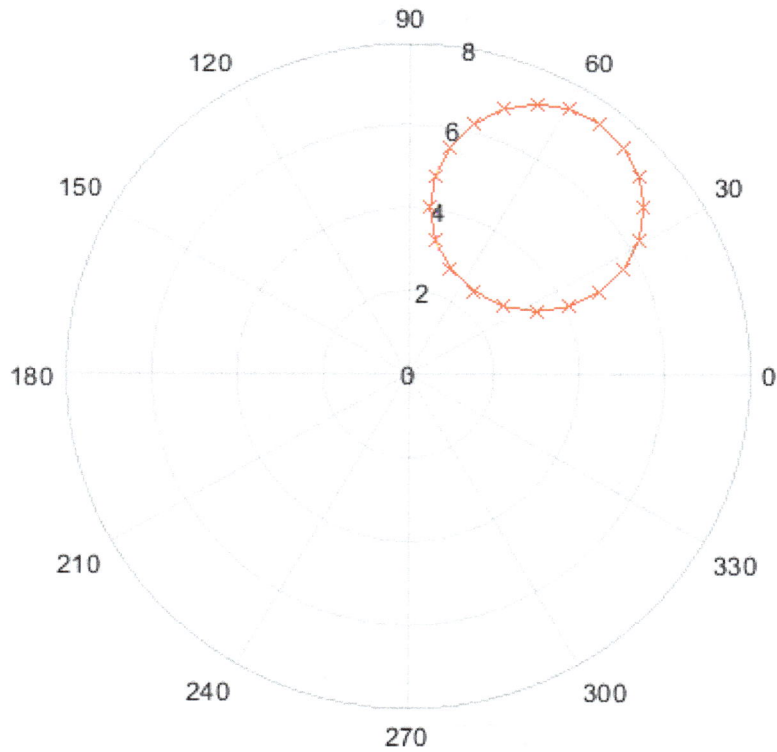

Figure-2.15

2.2 Multiplication and Division in Polar Form

2.2.1 Multiplication in Polar Form

Let
$$z_1 = r_1(\cos\theta_1 + i\sin\theta_1)$$
$$z_2 = r_2(\cos\theta_2 + i\sin\theta_2)$$

Then,
$$z_1 z_2 = r_1 r_2(\cos(\theta_1 + \theta_2) + i\sin(\theta_1 + \theta_2))$$

Proof-1

$$z_1 z_2 = r_1(\cos\theta_1 + i\sin\theta_1) r_2(\cos\theta_2 + i\sin\theta_2)$$

$$z_1 z_2 = r_1 r_2(\cos\theta_1 \cos\theta_2 + i\cos\theta_1 \sin\theta_2 + i\sin\theta_1 \cos\theta_2 - \sin\theta_1 \sin\theta_2)$$

$$z_1 z_2 = r_1 r_2(\cos\theta_1 \cos\theta_2 - \sin\theta_1 \sin\theta_2 + i(\sin\theta_1 \cos\theta_2 + \cos\theta_1 \sin\theta_2))$$

where using

$$\cos(\theta_1 + \theta_2) = \cos\theta_1 \cos\theta_2 - \sin\theta_1 \sin\theta_2$$
and
$$\sin(\theta_1 + \theta_2) = \sin\theta_1 \cos\theta_2 + \cos\theta_1 \sin\theta_2$$

we obtain

$$z_1 z_2 = r_1 r_2(\cos(\theta_1 + \theta_2) + i\sin(\theta_1 + \theta_2))$$

Proof-2:
$$z_1 = r_1(\cos\theta_1 + i\sin\theta_1)$$

can be written as
$$z_1 = r_1 e^{i\theta_1}$$

and
$$z_2 = r_2(\cos\theta_2 + i\sin\theta_2)$$

can be written as
$$z_2 = r_2 e^{i\theta_2}$$
Then,
$$z_1 z_2 = r_1 e^{i\theta_1} r_2 e^{i\theta_2} \rightarrow z_1 z_2 = r_1 r_2 e^{i(\theta_1 + \theta_2)}$$

where using the property

$$e^{i\theta} = \cos\theta + i\sin\theta$$
we get

$$z_1 z_2 = r_1 r_2 \big(\cos(\theta_1 + \theta_2) + i\sin(\theta_1 + \theta_2)\big)$$

In MCode-2.12 we illustrate the multiplication operation in polar coordinates using MATLAB.

MCode-2.12

```
r1 = 2;
theta1 = pi / 4;
% z1 = r1 * exp(j*theta1);
z1 = r1 * (cos(theta1)+ j*sin(theta1));

r2 = 4;
theta2 = pi / 2;
% z2 = r2 * exp(j*theta2);
z2 = r2 * (cos(theta2)+ j*sin(theta2));

abs(z1*z2) - (abs(z1)*abs(z2)) % = 0

angle(z1*z2) - (angle(z1)+angle(z2)) % = 0

polarplot(theta1,abs(z1),'r -x')
hold on;

polarplot(theta2,abs(z2),'r -x')

polarplot(theta1+theta2,abs(z1*z2),'r -*')
```

When MCode-2.12 is run, we get the graphic in Figure-2.16.

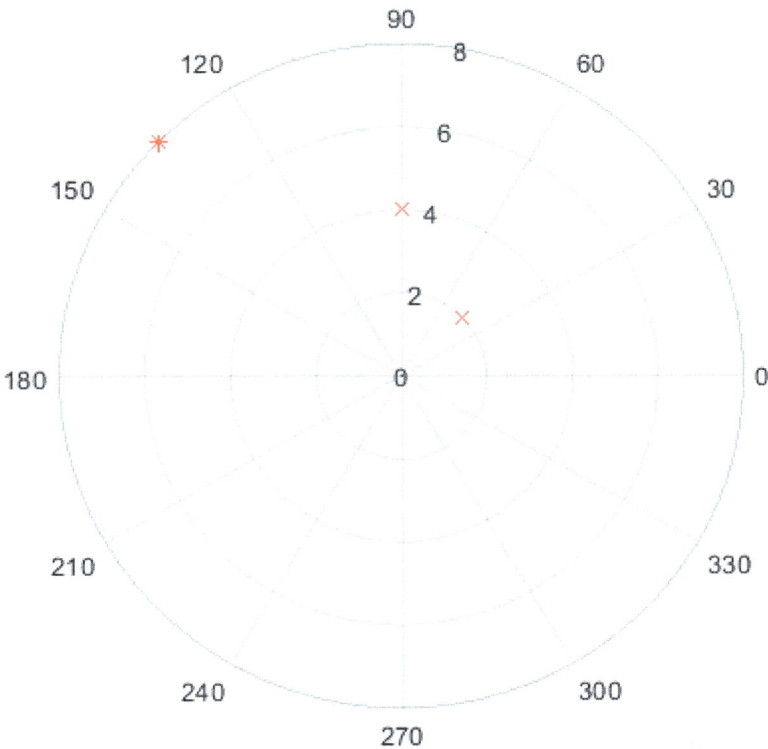

Figure-2.16

2.2.2 Division in Polar Form

Let
$z_1 = r_1(\cos\theta_1 + i\sin\theta_1)$

$z_2 = r_2(\cos\theta_2 + i\sin\theta_2)$
Then,

$$\frac{z_1}{z_2} = \frac{r_1}{r_2}(\cos(\theta_1 - \theta_2) + i\sin(\theta_1 - \theta_2))$$

Proof:
$z_1 = r_1(\cos\theta_1 + i\sin\theta_1)$

can be written as
$z_1 = r_1 e^{i\theta_1}$

and

$$z_2 = r_2(\cos\theta_2 + i\sin\theta_2)$$

can be written as
$$z_2 = r_2 e^{i\theta_2}$$
Then,
$$\frac{z_1}{z_2} = \frac{r_1}{r_2} e^{i(\theta_1 - \theta_2)}$$
where using the property

$$e^{i\theta} = \cos\theta + i\sin\theta$$
we get
$$\frac{z_1}{z_2} = \frac{r_1}{r_2}(\cos(\theta_1 - \theta_2) + i\sin(\theta_1 - \theta_2))$$

Using the obtained results, we can write

$$\arg z_1 z_2 = \arg z_1 + \arg z_2$$

$$\arg \frac{z_1}{z_2} = \arg z_1 - \arg z_2$$

In MCode-2.13 we illustrate the division operation in polar coordinates using MATLAB.

MCode-2.13

```
clc; clear all; close all;
r1 = 4;
theta1 = pi / 4;
% z1 = r1 * exp(j*theta1);
z1 = r1 * (cos(theta1)+ j*sin(theta1));

r2 = 2;
theta2 = pi / 2;
% z2 = r2 * exp(j*theta2);
z2 = r2 * (cos(theta2)+ j*sin(theta2));

abs(z1/z2) - (abs(z1)/abs(z2)) % = 0

angle(z1/z2) - (angle(z1)- angle(z2)) % = 0

polarplot(theta1,abs(z1),'r -x')
hold on;

polarplot(theta2,abs(z2),'r -x')
```

```
polarplot(theta1-theta2,abs(z1/z2),'r -*')
```

When MCode-2.13 is run, we get the graphic in Figure-2.17.

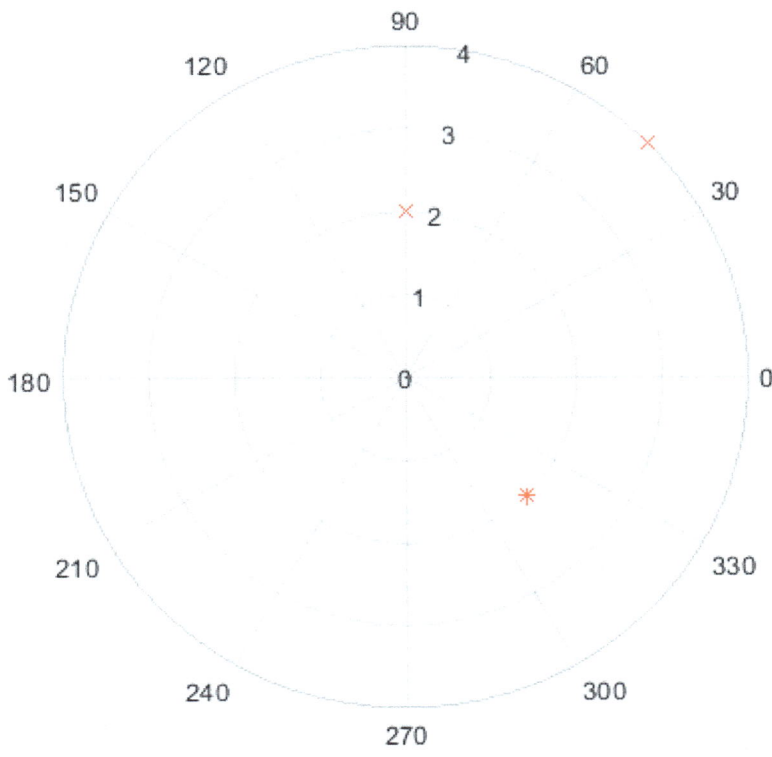

Figure-2.17

2.2.3 de Moivre's formula

The formula below is called de Moivre's formula

$$(\cos \theta + i\sin \theta)^n = \cos n\theta + i\sin n\theta$$

If we have

$$z_1 z_2 = r_1 r_2 (\cos (\theta_1 + \theta_2) + i\sin (\theta_1 + \theta_2))$$

then for
$$z^2 = r^2(\cos 2\theta + i\sin 2\theta)$$

and in general we can write

$$z^n = r^n(\cos n\theta + i\sin n\theta)$$

Example-2.4:

$z = r(\cos\theta + i\sin\theta)$

Find z^{-1}

Solution-2.4: The complex number z can be written as

$z = re^{i\theta}$

Then,

$z^{-1} = \frac{1}{r}e^{-i\theta}$

$z^{-1} = \frac{1}{r}(\cos(-\theta) + i\sin(-\theta))$

leading to

$z^{-1} = \frac{1}{r}(\cos(\theta) - i\sin(\theta))$

In MCode-2.14 we illustrate power operation in polar coordinates using MATLAB.

MCode-2.14

```
% z = r(cos(θ)+j sin(θ))
% z^n = r^n(cos(nθ)+j sin(nθ))

r = 1;
theta = pi / 4;
% z = r * exp(j*theta);
z = r * (cos(theta)+ j*sin(theta));

z5 = z^5;
w5 = r^5 * (cos(5*theta)+ j*sin(5*theta));

z5-w5   % = 0

polarplot(z,'r -x')
hold on;

polarplot(z5,'b -*')
```

```
% polarplot(w5,'g -*')
```

When MCode-2.14 is run, we get the graphic in Figure-2.18.

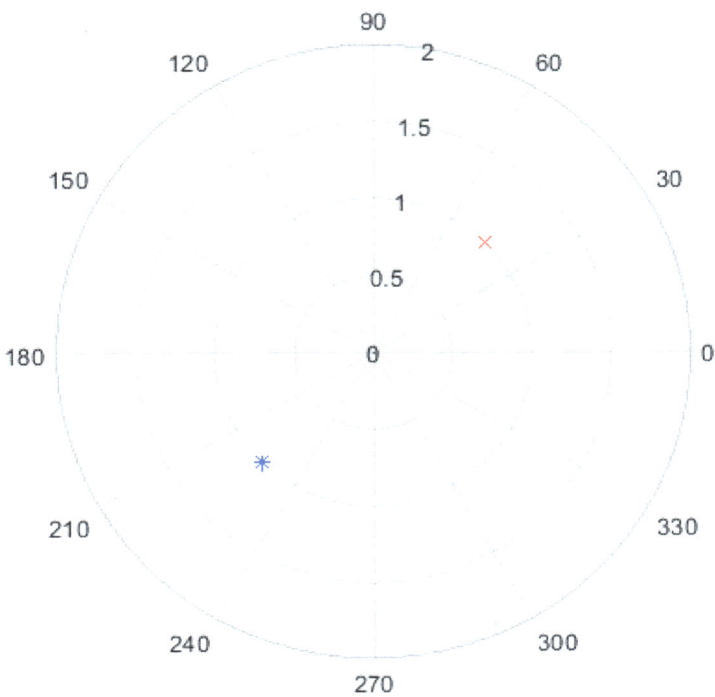

Figure-2.18

Example-2.5: Calculate
$(1 + i)^7$

Solution-2.5: The complex number

$z = 1 + i$
has the polar form
$$z = \sqrt{2}\left(\cos\frac{\pi}{4} + i\sin\frac{\pi}{4}\right)$$

Then, we have
$$z^7 = (\sqrt{2})^7\left(\cos\frac{7\pi}{4} + i\sin\frac{7\pi}{4}\right) \rightarrow$$

where
$$(\sqrt{2})^7 = 2^{\frac{7}{2}} \rightarrow (\sqrt{2})^7 = 2^{\frac{6}{2}}2^{\frac{1}{2}} \rightarrow (\sqrt{2})^7 = 8\sqrt{2}$$

and
$$\cos\frac{7\pi}{4} = \cos\left(\frac{7\pi}{4} - 2\pi\right) \rightarrow \cos\left(\frac{\pi}{4}\right) = \frac{1}{\sqrt{2}}$$

$$\sin\frac{7\pi}{4} = \sin\left(\frac{7\pi}{4} - 2\pi\right) \rightarrow \sin\left(-\frac{\pi}{4}\right) = -\frac{1}{\sqrt{2}}$$

Then, we get
$$z^7 = 8\sqrt{2}\left(\frac{1}{\sqrt{2}} - i\frac{1}{\sqrt{2}}\right)$$

$$z^7 = 8 - i8$$

$$z^7 = 8(1 - i)$$

Exercise: Given
$z = -1 + i$

Find z^8

In MCode-2.15 we illustrate the power calculation both in rectangular and polar coordinates using MATLAB.

MCode-2.15

```
z = complex(1,1); % z = 1+j

r = abs(z);   % 1.4141

theta = angle(z); % pi / 4 = 0.7854

z7 = z^7;

w7 = r^7 * (cos(7*theta) + j*sin(7*theta));

z7-w7 % = 0

polarplot(z,'r *')
hold on;

polarplot(z7,'b *')
```

When MCode-2.15 is run, we get the graphic in Figure-2.19.

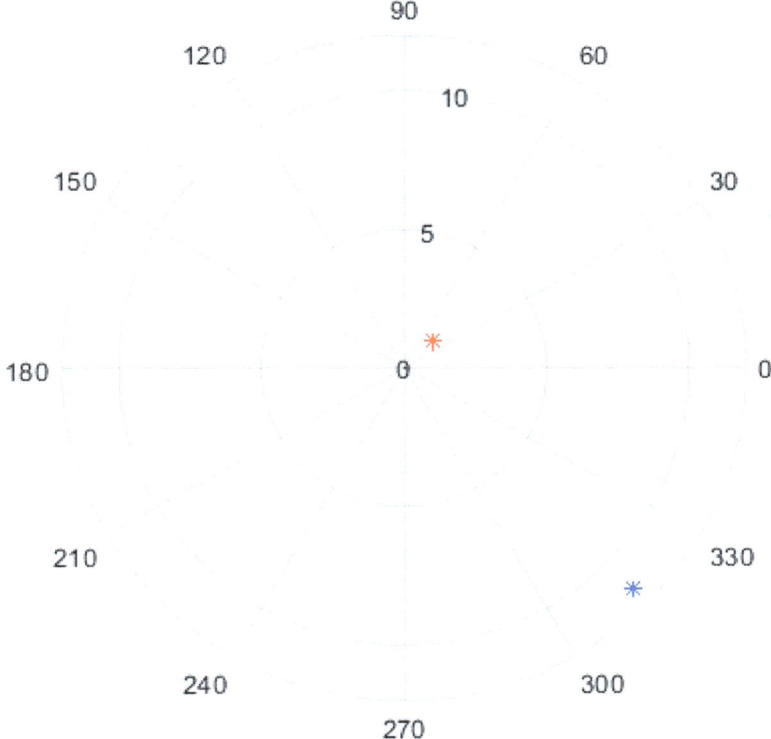

Figure-2.19

Example-2.6: Calculate the argument of

$$z = \frac{i}{-1-i}$$

Solution-2.6:

$\arg z = \arg i - \arg(-1-i)$

$\text{Arg } i = \frac{\pi}{2}$

$\text{Arg}(-1-i) = -\frac{3\pi}{4}$

$\frac{\pi}{2} - \left(-\frac{3\pi}{4}\right) = \frac{5\pi}{4}$

Since the principal argument should be in the interval $[-\pi..\pi)$, we can add -2π to the result obtaining

$$\frac{5\pi}{4} - 2\pi = -\frac{3\pi}{4}$$

then we have,

$$\arg z = -\frac{3\pi}{4} + k2\pi$$

Note that $\arg(z_1/z_2) = \arg z_1 - \arg z_2$

But
$\text{Arg}(z_1/z_2) \neq \text{Arg } z_1 - \text{Arg } z_2$

2.3 Roots of a Complex Number

Let
$z = r(\cos\theta + i\sin\theta)$

$w = R(\cos\phi + i\sin\phi)$

If
$w^n = z$

there are n roots of z denoted by w

To find the roots we first write z as

$z = r[\cos(\theta + k2\pi) + i(\theta + k2\pi)]$

next we write
$w = z^{\frac{1}{n}}$
from which we get,

$$z = r^{\frac{1}{n}}\left[\cos\left(\frac{\theta + k2\pi}{n}\right) + i\left(\frac{\theta + k2\pi}{n}\right)\right]$$

which can be written as

$$z = r^{\frac{1}{n}}\exp\left(\frac{\theta + k2\pi}{n}\right)$$

and the roots are obtained giving values to k as

$k = 0, 1, 2, ..., n - 1$
These n values are the roots of z

Example-2.7: Given $z = 1$, calculate

$\sqrt[3]{1} \quad \sqrt[4]{1} \quad \sqrt[6]{1}$

Solution-2.7:
$z = 1 \rightarrow z = 1(\cos 0 + i\sin 0)$

$z = 1(\cos(0 + k2\pi) + i\sin(0 + k2\pi))$

$z = \cos k2\pi + i\sin k2\pi$

Let
$w^n = z$
then we have,
$w = \cos k\dfrac{2\pi}{n} + i\sin k\dfrac{2\pi}{n} \quad k = 0, 1, \ldots n-1$

Roots for $\sqrt[3]{1}$ are found as:

$w = \cos k\dfrac{2\pi}{3} + i\sin k\dfrac{2\pi}{3} \quad k = 0, 1, 2$

For $k = 0$
$w_0 = \cos 0 + i\sin 0 \rightarrow w_0 = 1$

For $k = 1$
$w_1 = \cos \dfrac{2\pi}{3} + i\sin \dfrac{2\pi}{3} \rightarrow w_1 = -\dfrac{1}{2} + \dfrac{\sqrt{3}}{2}i$

For $k = 2$
$w_2 = \cos \dfrac{4\pi}{3} + i\sin \dfrac{4\pi}{3} \rightarrow w_2 = -\dfrac{1}{2} - \dfrac{\sqrt{3}}{2}i$

Locations of the roots for $\sqrt[3]{1}$ are shown in the Figure-2.20.

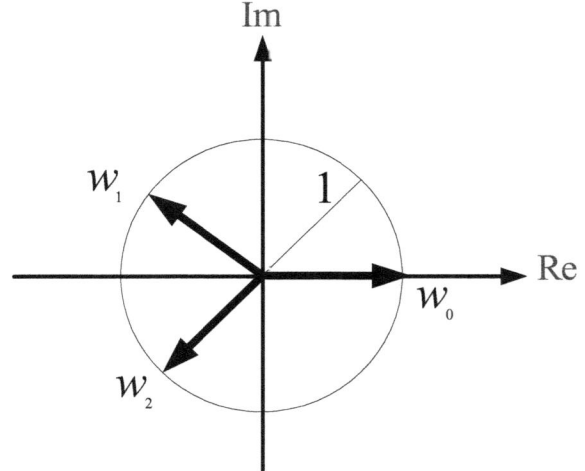

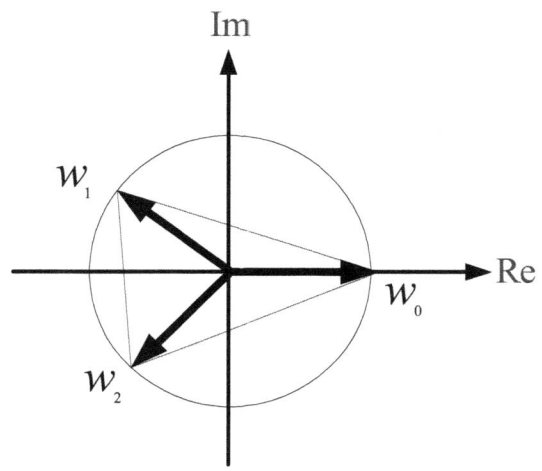

Figure-2.20

Principal arguments of the roots are show below

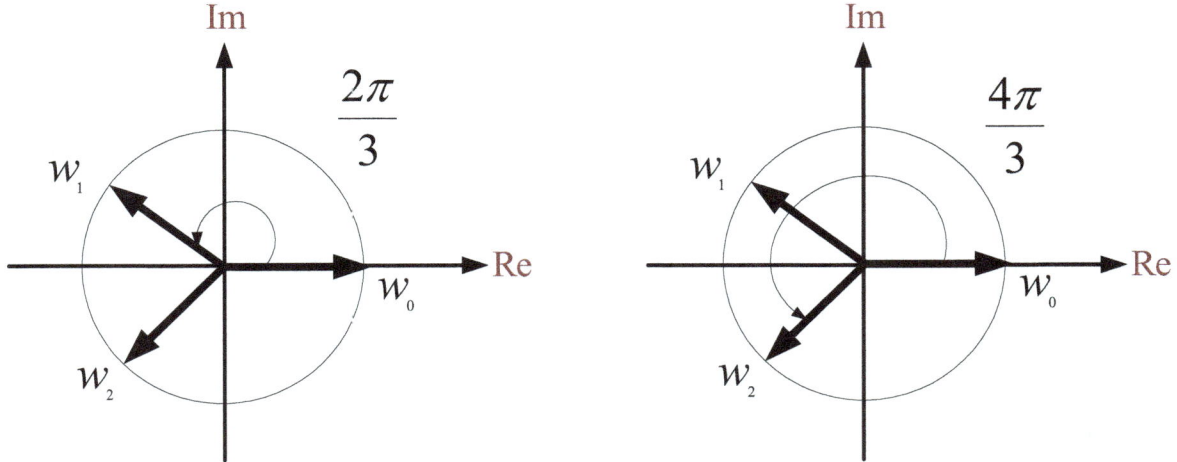

Figure-2.21

Note that the roots are separated with equal angles. In MCode-2.16, we calculate the roots by MATLAB.

MCode-2.16

```
% z^3 - 1 = 0

p = [1 0 0 -1]  % z^3 - 1 = 0

z = roots(p)
% poly2sym(p,'z')    % check to see if you have the right polynomial with the poly2sym

polarplot(z,'r *')

% zvect(z)
```

When MCode-2.16 is run, we get the graphic in Figure-2.22.

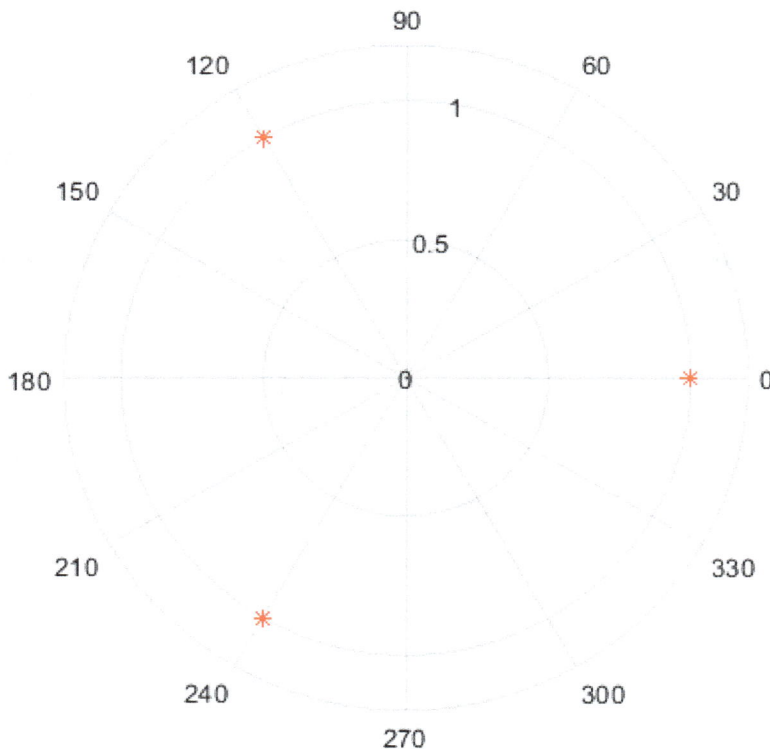

Figure-2.22

Roots for $\sqrt[4]{1}$ are found as:

$$w = \cos k\frac{2\pi}{4} + i\sin k\frac{2\pi}{4} \quad k = 0, 1, 2, 3$$

For $k = 0$
$w_0 = \cos 0 + i\sin 0 \quad \rightarrow \quad w_0 = 1$

For $k = 1$
$w_1 = \cos \frac{2\pi}{4} + i\sin \frac{2\pi}{4} \rightarrow w_1 = i$

For $k = 2$
$w_2 = \cos \frac{4\pi}{4} + i\sin \frac{4\pi}{4} \rightarrow w_2 = -1$

For $k = 3$

$$w_3 = \cos\frac{6\pi}{4} + i\sin\frac{6\pi}{4} \rightarrow w_3 = -i$$

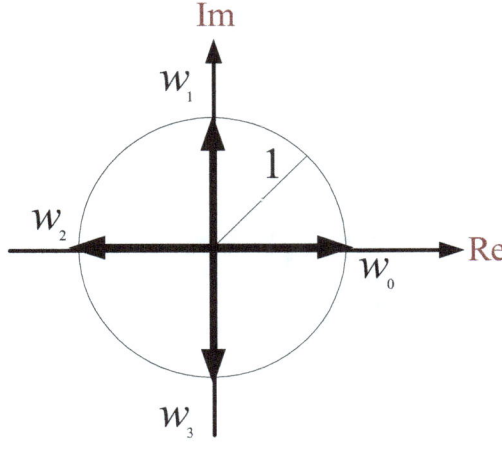

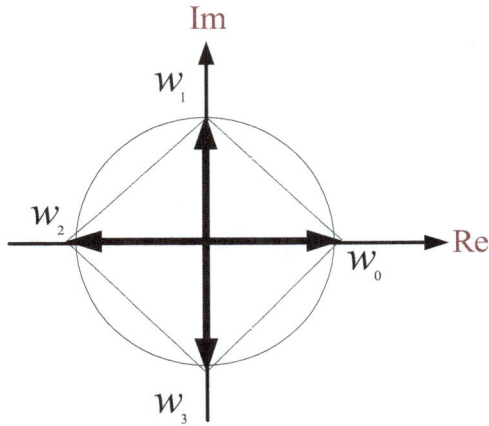

Figure-2.23

In MCode-2.17 we calculate the roots of $\sqrt[4]{1}$ in MATLAB.

MCode-2.17

```
% z^4 - 1 = 0

p = [1 0 0 0 -1] % z^4 - 1 = 0

z = roots(p)
% poly2sym(p,'z')   % check to see if you have the right
polynomial with the poly2sym

% zprint(z)

polarplot(z,'r *')
```

When MCode-2.17 is run, we get the graphic in Figure-2.24.

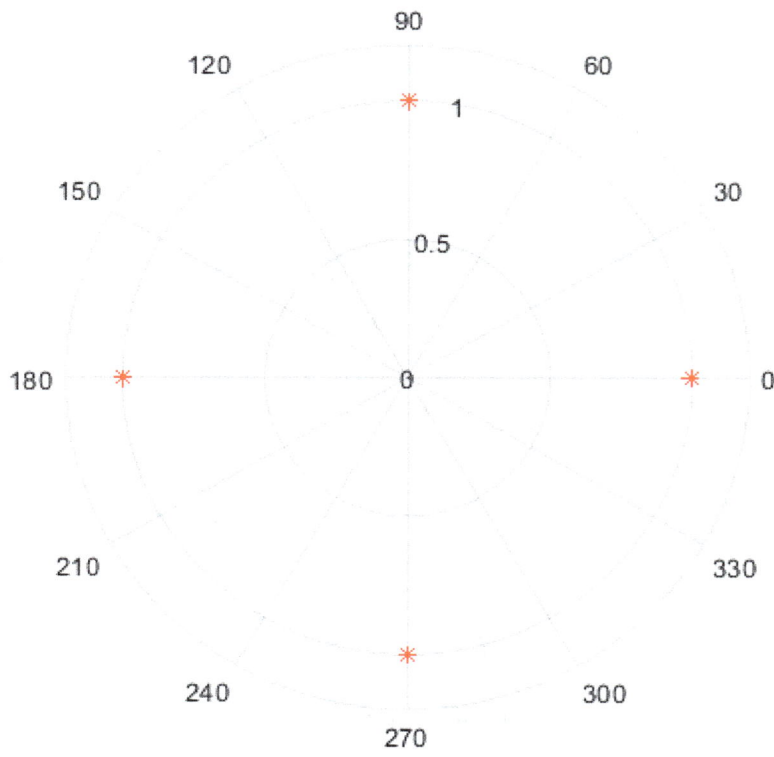

Figure-2.24

In MCode-2.17 we calculate the roots of $\sqrt[5]{1}$ in MATLAB.

MCode-2.18

```
% z^5 - 1 = 0

p = [1 0 0 0 0 -1]   % z^5 - 1 = 0

z = roots(p)

polarplot(z,'r *')

figure;

zvect(z)
```

When MCode-2.18 is run, we get the graphic in Figure-2.25.

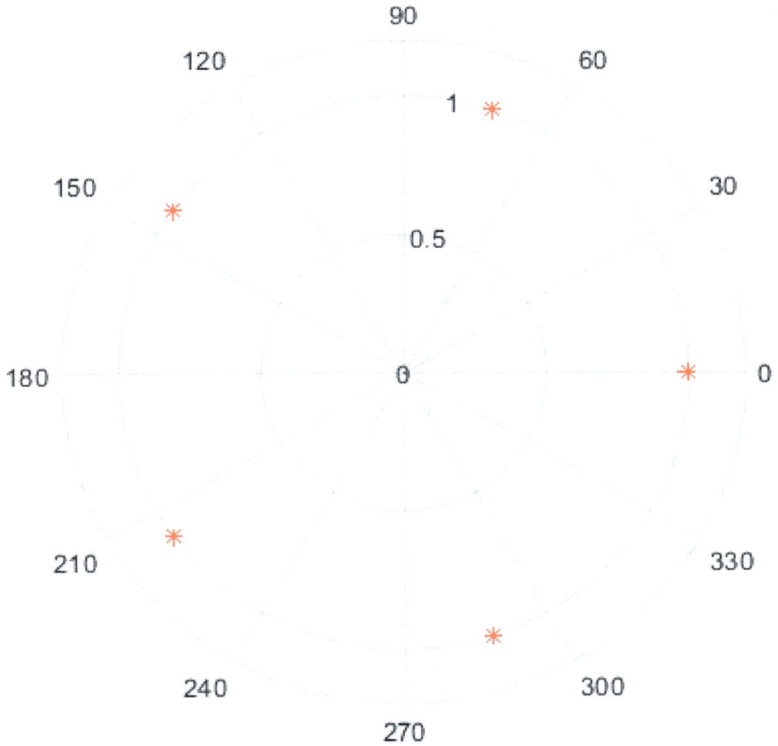

Figure-2.25

The roots of $\sqrt[5]{1}$ in are displayed using vectors in Figure-2.26.

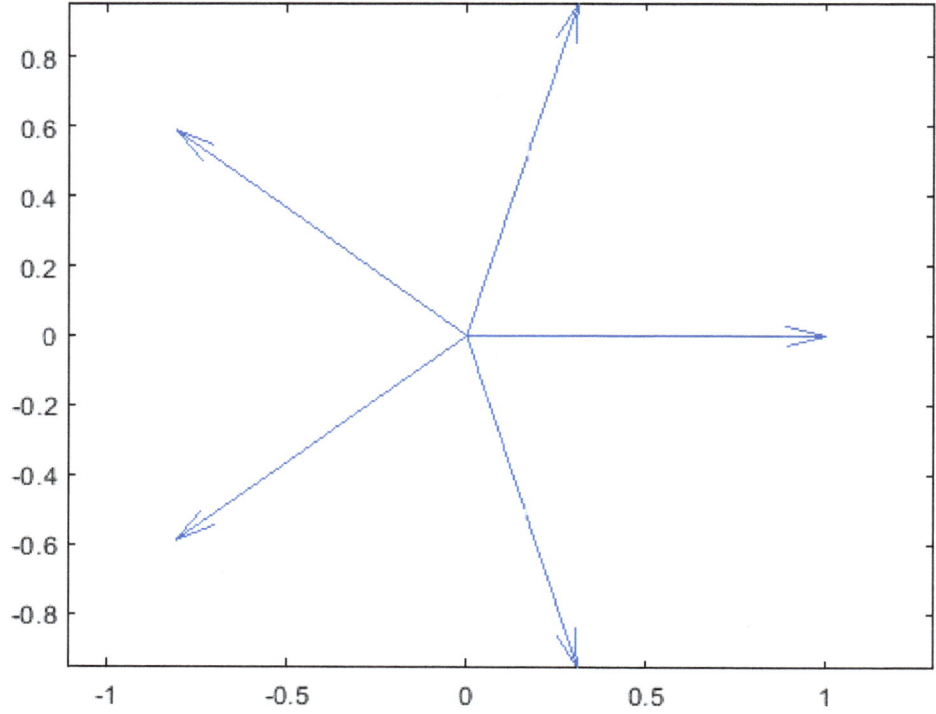

Figure-2.26

Chapter-3

Introduction to Complex Numbers

Abstract: In this chapter we explain complex plane and give some fundamental definitions such as open set, closed set, connected set, domain, contour, etc. We provide comprehensive MATLAB examples for specifiying complex regions. We also explain complex line integration both theoretically and by MATLAB examples as well.

3.1 Complex Plane

Every complex number is indicated by a single dot in the complex plane which is a rectangular plane. The horizontal axis of the complex plane is used for the real part of the complex numbers and the vertical axis of the complex plane is used for the imaginary part of the complex numbers. In the complex plane of Figure-3.1 a single complex number is displayed.

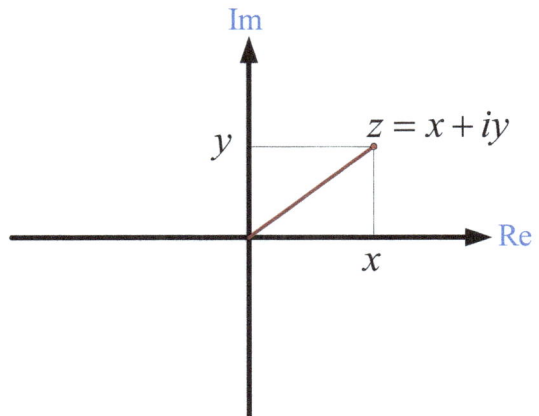

Figure-3.1

In MCode-3.1 we display a number of complex numbers in a complex plane.

MCode-3.1

```matlab
% Number of complex numbers in a row of complex plane:
N = 10;

real_axis = linspace(-2, 2, N);

imag_axis = linspace(-2, 2, N);

[x, y] = meshgrid(real_axis, imag_axis);

% Generate a grid of complex numbers:

z = x + j * y;

plot(real(z), imag(z), 'k .')
axis([-2.5 2.5 -2.5 2.5]);
xlabel('Real Axis');
ylabel('Imaginary Axis');
title('Complex Number Plane');
```

When MCode-3.1 is run, we get the graph in Figure-3.2.

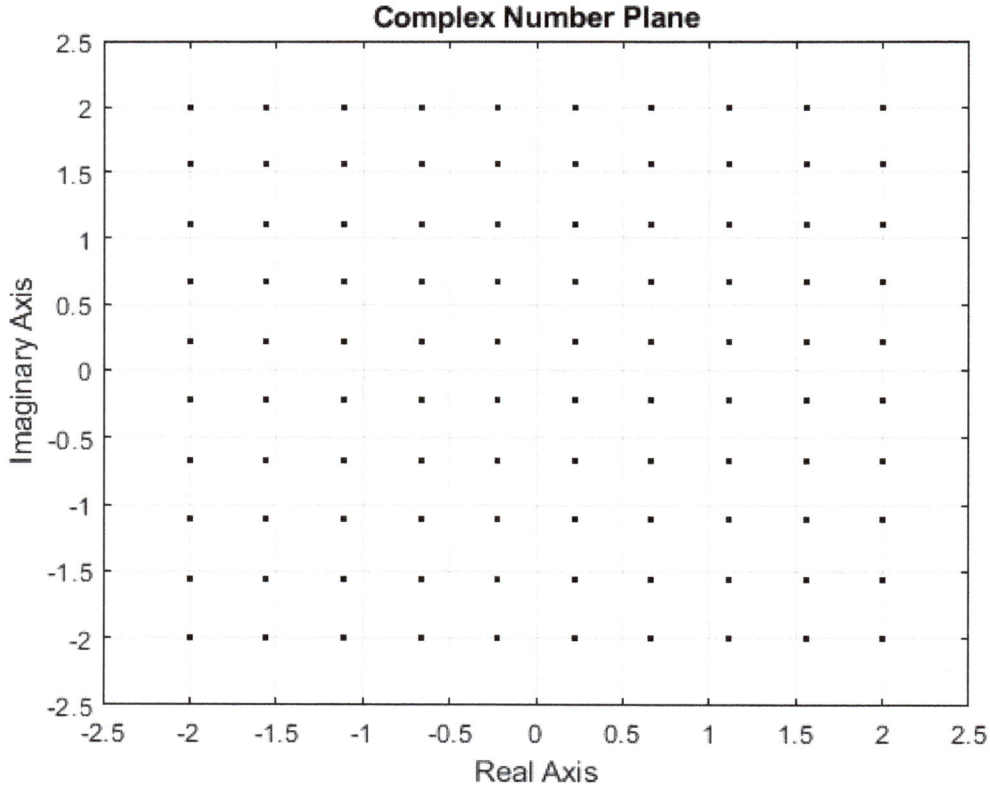

Figure-3.2

3.2 Regions in the Complex Plane

ϵ neighborhood of the complex number z_0 is defined as

$$|z - z_0| < \epsilon$$

The graph of

$$0 < |z - z_0| < \epsilon$$

can be drawn as in Figure-3.3.

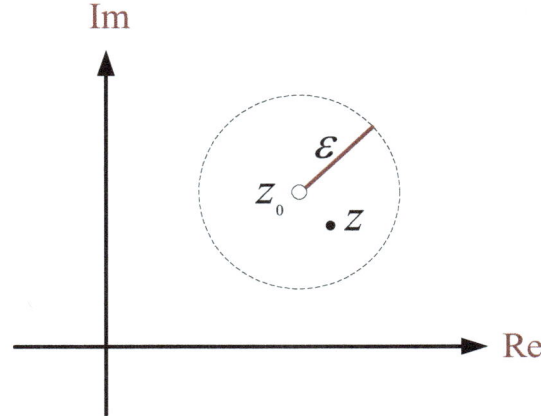

Figure-3.3

The graph of

$$0 \leq |z - z_0| \leq \epsilon$$

can be drawn as in Figure-3.4.

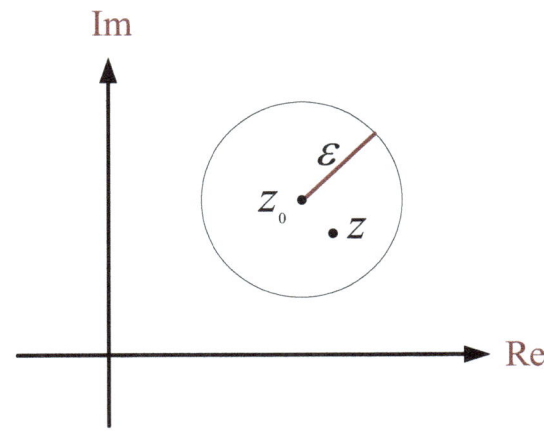

Figure-3.4

The graph of
$|z - 1| < 0.5$

is drawn using MATLAB in MCode-3.2.

MCode-3.2

```matlab
% Number of complex numbers in a row of complex plane:
N = 17;

real_axis = linspace(-2, 2, N);

imag_axis = linspace(-2, 2, N);

[x, y] = meshgrid(real_axis, imag_axis);

% Generate a grid of complex numbers:

z = x + j * y;

plot(real(z), imag(z), 'k .')
axis([-2.5 2.5 -2.5 2.5]);
xlabel('Real Axis');
ylabel('Imaginary Axis');
title('Complex Number Plane');
grid on;

c = abs(z-1) <= 0.5;

hold on;
plot(real(z(c)), imag(z(c)), 'r *')

title('Complex plane set |z-1| < 0.5')
```

The When MCode-3.2 is run, we get the graph in Figure-3.5.

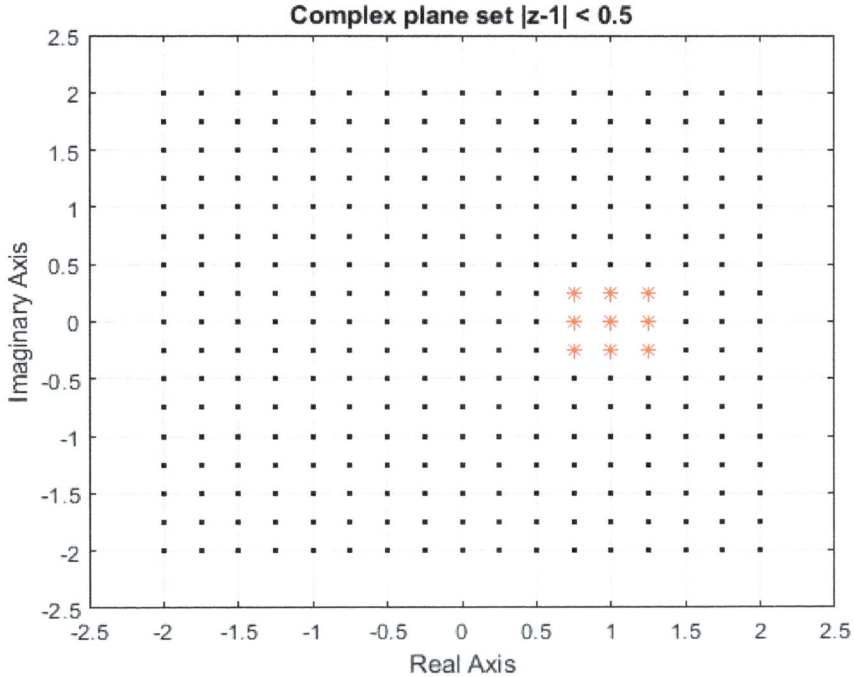

Figure-3.5

The graph of $|z - 1| \leq 0.5$ is shown in Figure-3.6.

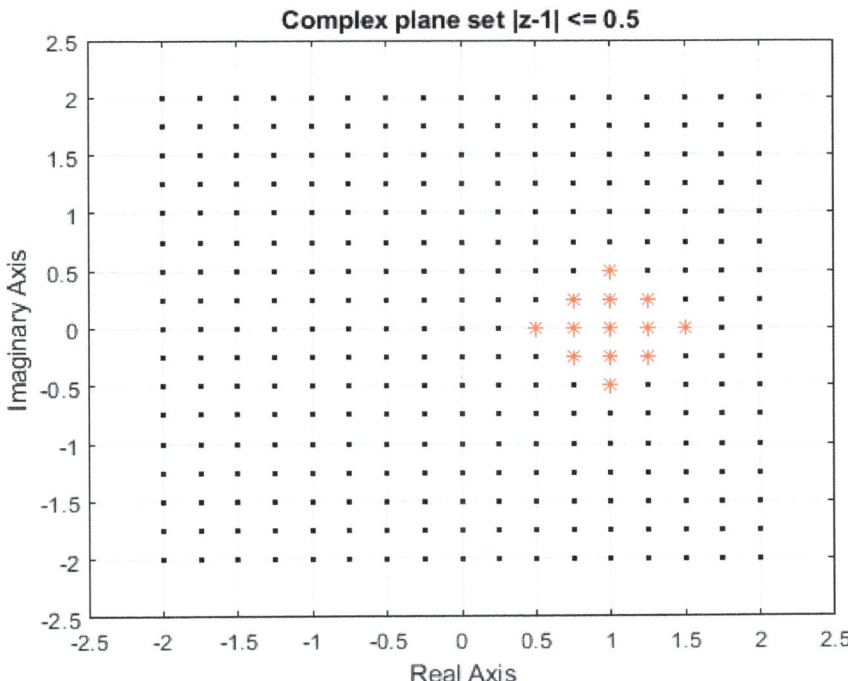

Figure-3.6

The graph of $|z - 1| < 0.8$ can be drawn using MCode-3.3.

MCode-3.3

```
% Number of complex numbers in a row of complex plane:
N = 10;

real_axis = linspace(-2, 2, N);

imag_axis = linspace(-2, 2, N);

[x, y] = meshgrid(real_axis, imag_axis);

% Generate a grid of complex numbers:

z = x + j * y;

plot(real(z), imag(z), 'k .')
axis([-2.5 2.5 -2.5 2.5]);
xlabel('Real Axis');
ylabel('Imaginary Axis');
title('Complex Number Plane');
grid on;

c = abs(z-1) < 0.8;

hold on;
plot(real(z(c)), imag(z(c)), 'r *')

title('Complex plane set |z-1| < 0.8')
```

When MCode-3.3 is run, we get the graph in Figure-3.7.

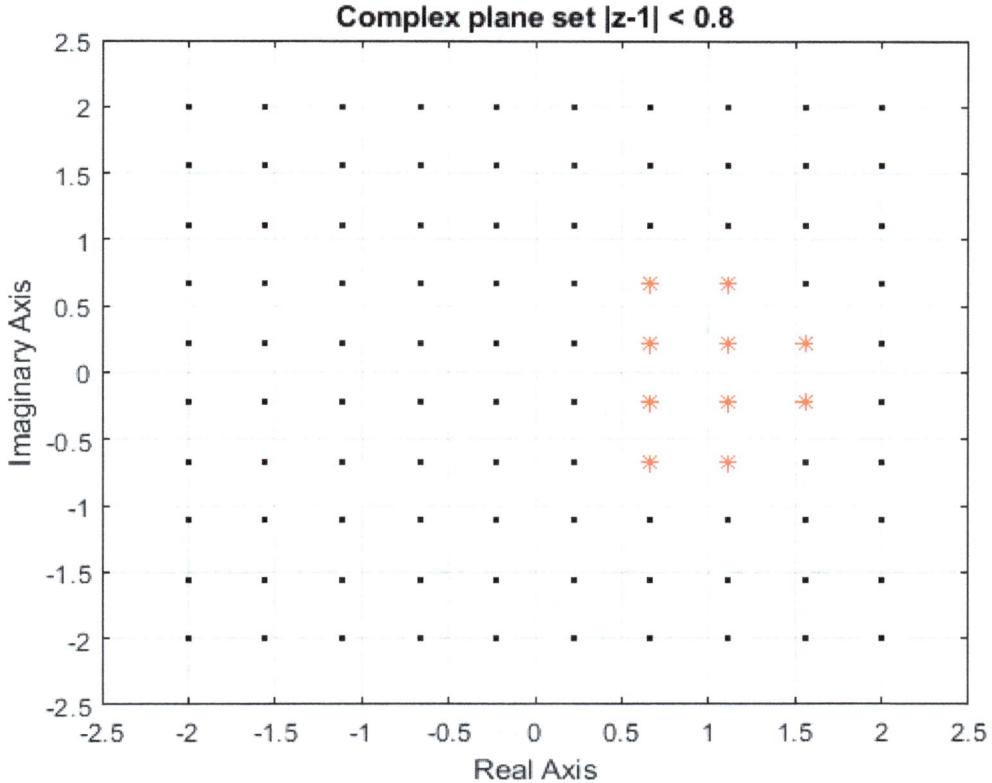

Figure-3.7

We can increase the number of the points in the complex plane and draw the graph of as in MCode-3.4.

MCode-3.4

```
% Number of complex numbers in a row of complex plane:
N = 50;

real_axis = linspace(-2, 2, N);

imag_axis = linspace(-2, 2, N);

[x, y] = meshgrid(real_axis, imag_axis);

% Generate a grid of complex numbers:

z = x + j * y;

plot(real(z), imag(z), 'k .')
axis([-2.5 2.5 -2.5 2.5]);
xlabel('Real Axis');
ylabel('Imaginary Axis');
```

```
title('Complex Number Plane');
grid on;

c = abs(z-1) < 0.8;

hold on;
plot(real(z(c)), imag(z(c)), 'r *')

title('Complex plane set |z-1| < 0.8')
```

When MCode-3.4 is run, we get the graph in Figure-3.8..

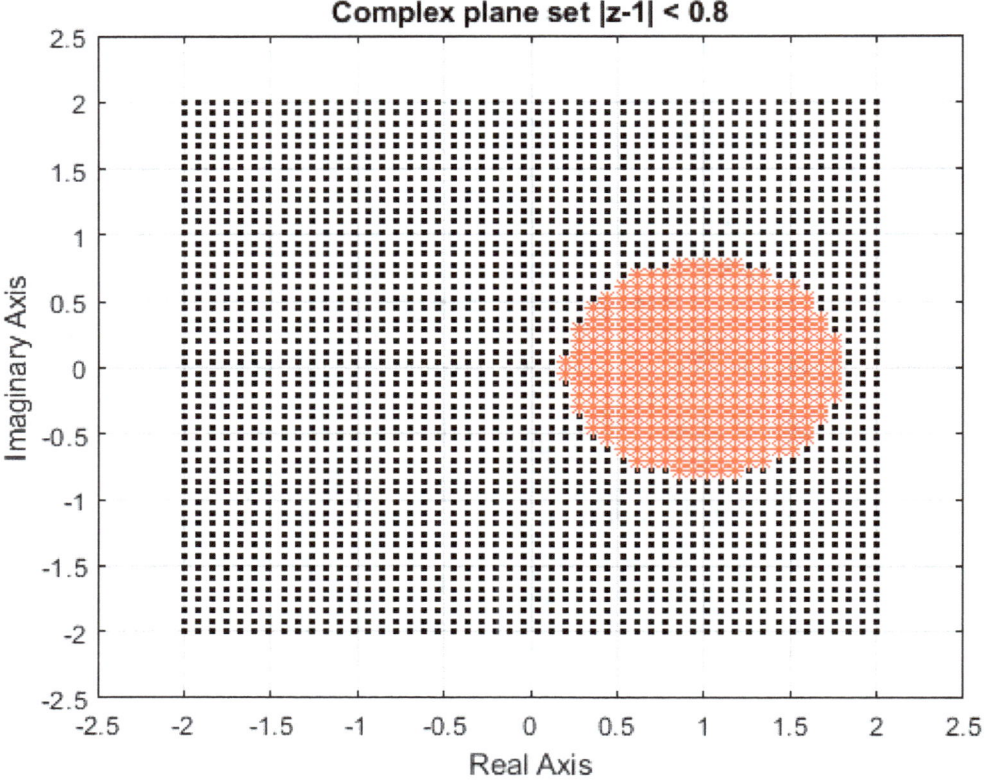

Figure-3.8

Example-3.1: The graph of
$|1 + z + z^2| < 2.5$
can be drawn using MCode-3.5.

MCode-3.5

```
% Number of complex numbers in a row of complex plane:
N = 50;
```

```
real_axis = linspace(-4, 4, N);
imag_axis = linspace(-4, 4, N);
[x, y] = meshgrid(real_axis, imag_axis);

% Generate a grid of complex numbers:
z = x + j * y;

plot(real(z), imag(z), 'k .')
xlabel('Real Axis');
ylabel('Imaginary Axis');
title('Complex Number Plane');
grid on;

c = abs(1+z+z.^2) <= 2.5;

hold on;
plot(real(z(c)), imag(z(c)), 'r *')

title('Complex plane set |1+z+z^2| <= 2.5')
```

When MCode-3.4 is run, we get the graph in Figure-3.9..

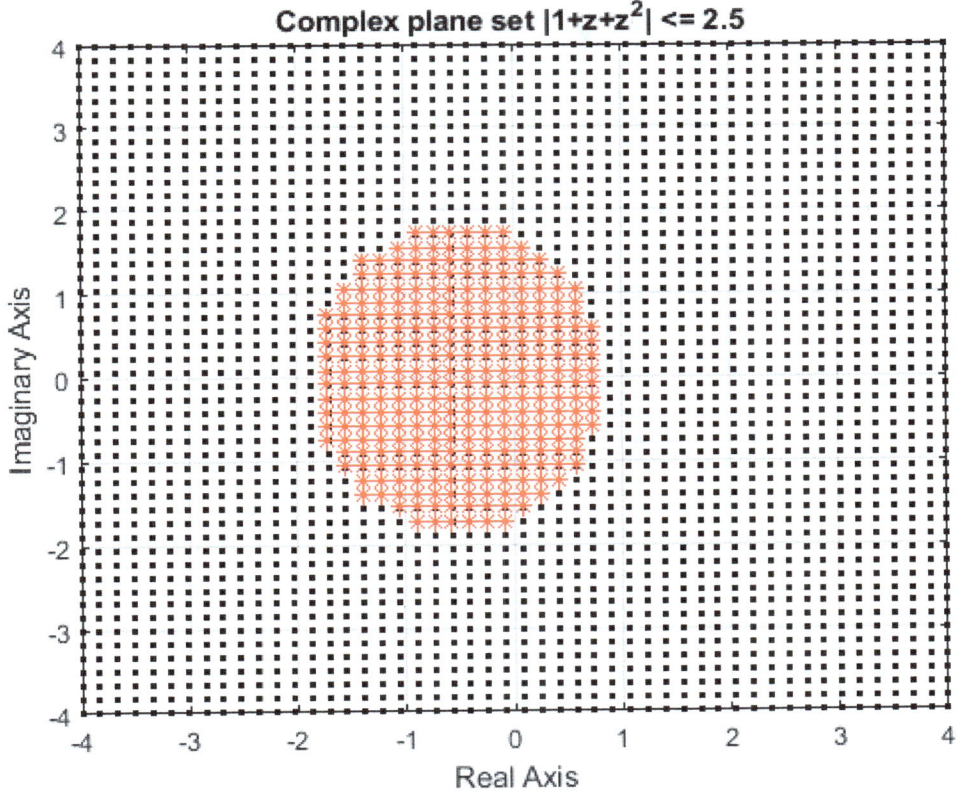

Figure-3.9

3.3 Definitions for Complex Regions

Interior Point

A point z_0 is an **interior point** of the complex set S if any neighborhood of z_0 contains only points of S

Exterior Point

A point z_0 is an **exterior point** of the complex set S if we can find a neighborhood of z_0 that does not contain a single point of S.

Boundary Point

If z_0 is neither an interior point nor an exterior point, then it is a **boundary point** of S.

The neighborhood of a **boundary point** contains at least one point in S and at least one point not in S.

Boundary

The set of all boundary points is called the **boundary** of S.

Properties of a Set

A set can be open, closed, and connected.

Open Set

An open set does not contain any of its boundary points.

Every point of an open is an interior point. The graph of an open set is depicted in Figure-3.10.

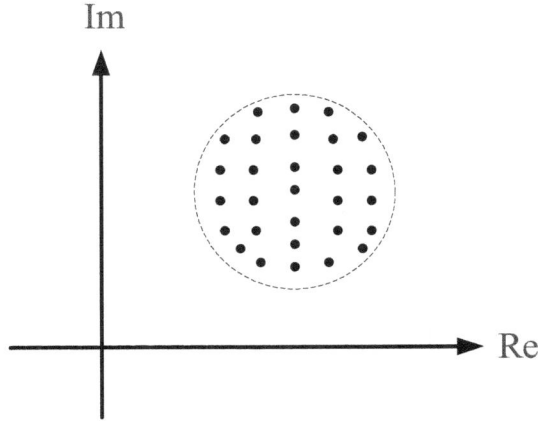

Figure-3.10 An open set

Closed Set

A closed set contains all of its boundary points The graph of a closed set is depicted in Figure-3.11.

Closure of a Set

Closure is defined for a closed set, and it is the set containing all boundary and interior points.

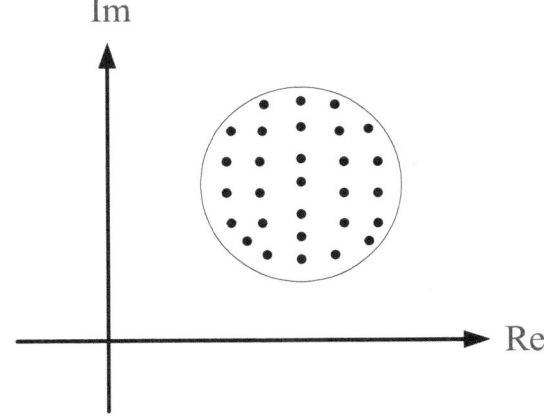

Figure-3.11 A closed set

Example-3.2:

An open set
$|z| < 1.5$

Example-3.3:

Closed set

$|z| \leq 1.5$

There can be some sets which are neither open nor closed.

If even a single boundary point is included in a set, then the set is not an open set, and if even a single boundary point is not included in a set, then the set is not a closed set.

Example-3.4: The disk expressed using
$0 < |z| \leq 2$
is neither an open nor a closed set.

Example-3.5: The set of all complex numbers is both open and closed since it has no boundary points.

Connected Set

If two points z_1 and z_2 of an open set can be connected by a finite number of line segments as depicted in Figure-3.12, then the set is a **connected** set, or we simply say that is is connected.

An open set S is **connected** if each pair of points z_1 and z_2 in it can be joined by a number of lines that lies entirely in S.

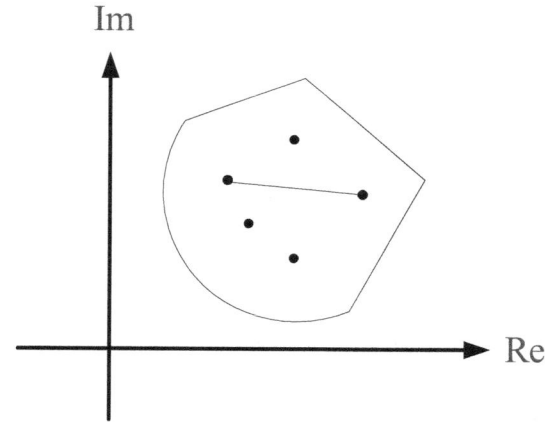

Figure-3.12 A connected set

Example-3.6: The open set $|z| < 1$ is a connected set.

Example-3.7: The annulus $1 < |z| < 2$ is connected. The graph of the annulus is depicted in Figure-3.13.

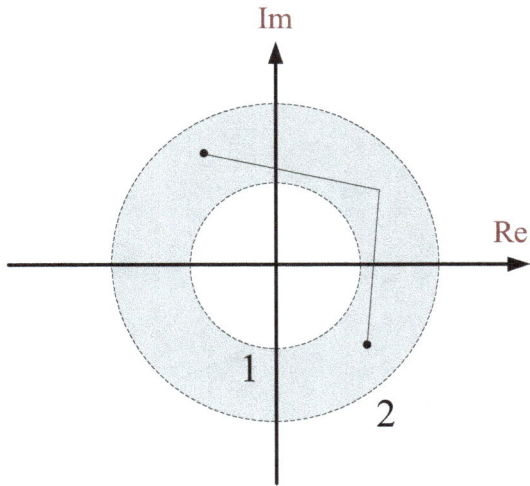

Figure-3.13

Domain

Aconnected open set is called a **domain**.

Region

A region is nothing but a domain with some or all of its boundary points.

Bounded Set

If we can draw a circle around a set, then the set if bounded, otherwise it is unbounded.

Example-3.8: The sets

$$|z| < 2.4$$

and

$$|z| \leq 2.4$$

are bounded sets

In MCode-3.6 the bounded set $|z| \leq 2.4$ is drawn.

MCode-3.6

```
% Number of complex numbers in a row of complex plane:
N = 50;

real_axis = linspace(-3, 3, N);
imag_axis = linspace(-3, 3, N);

[x, y] = meshgrid(real_axis, imag_axis);
```

```
% Generate a grid of complex numbers
z = x + j * y;

plot(real(z), imag(z), 'k .')
xlabel('Real Axis');
ylabel('Imaginary Axis');
title('Complex Number Plane');
grid on;

c = abs(z) <= 2.4 ;

hold on;
plot(real(z(c)), imag(z(c)), 'r *')
% axis([-1 1 -1.1 1])
title('Complex plane set for |z|<=2.4')
```

When MCode-3.6 is run, we get the graph in Figure-3.14.

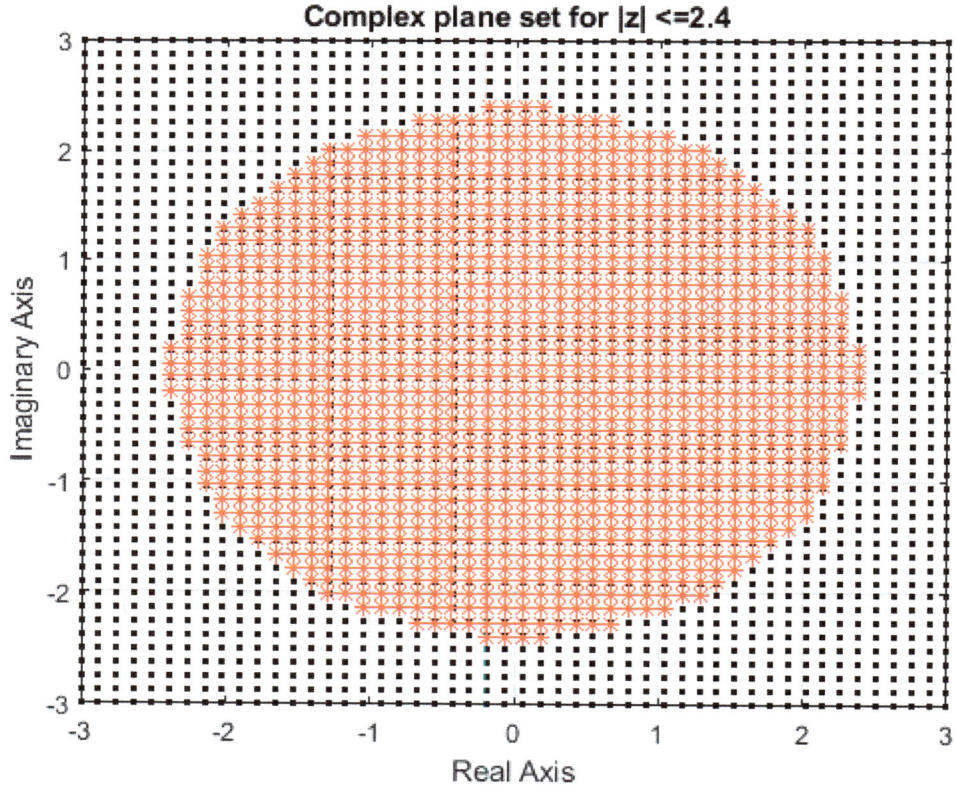

Figure-3.14

Example-3.9: The set indicated by $Re\{z\} \geq 0$

is unbounded. The graph of $Re\{z\} \geq 0$

can be plotted using MATLAB as in Figure-3.15.

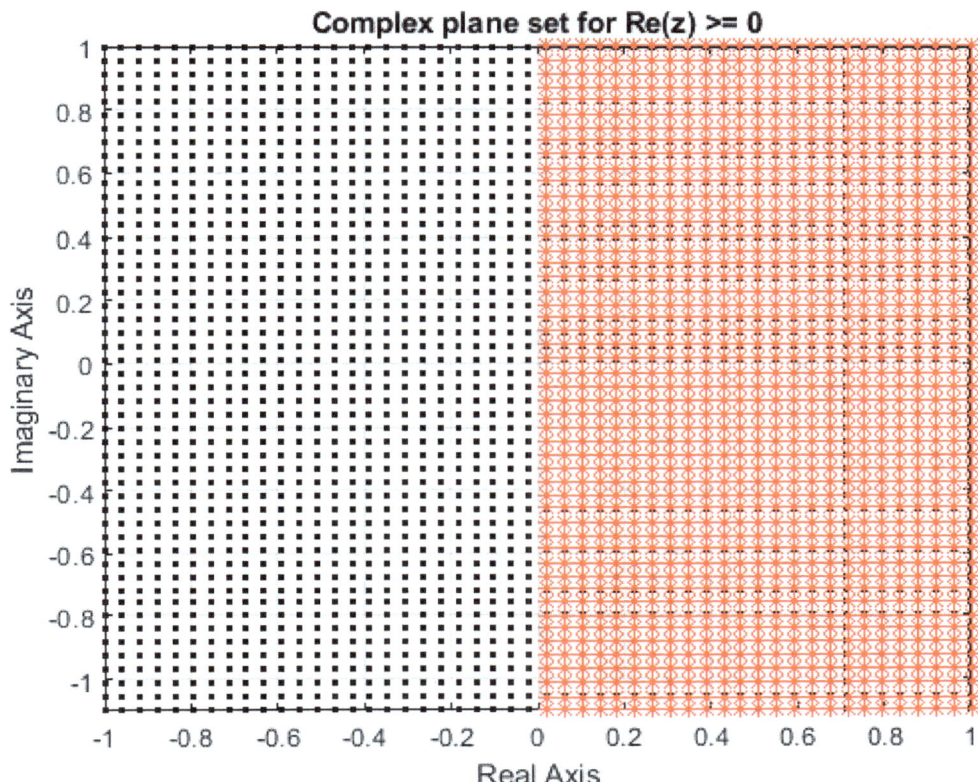

Figure-3.15

Example-3.10: Sketch the set

$$Im\left(\frac{1}{z}\right) > 1 \tag{3.1}$$

Solution:
$$\frac{1}{z} = \frac{1}{x+iy} \rightarrow \frac{1}{z} = \frac{x-iy}{(x+iy)(x-iy)} \rightarrow \frac{1}{z} = \frac{x-iy}{x^2+y^2}$$

Then,
$$Im\left(\frac{x-iy}{x^2+y^2}\right) > 1 \rightarrow -\frac{y}{x^2+y^2} > 1$$

leading to

$$x^2 + y^2 + y < 0$$

which can be written as

$$x^2 + \left(y^2 + y + \frac{1}{4}\right) < \frac{1}{4} \quad \rightarrow$$

$$(x - 0)^2 + \left(y + \frac{1}{2}\right)^2 = \left(\frac{1}{2}\right)^2$$

So inequality (3.1) represents the region interior to the circle as shown in Figure-3.16.

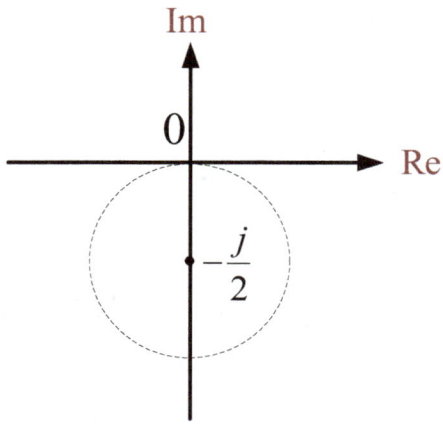

Figure-3.16

The graph of

$$Im\left(\frac{1}{z}\right) > 1$$

can be drawn using the MATLAB code in MCode-3.7.

MCode-3.7

```
% Number of complex numbers in a row of complex plane:
N = 50;

real_axis = linspace(-1, 1, N);
imag_axis = linspace(-1.1, 1, N);
[x, y] = meshgrid(real_axis, imag_axis);

% Generate a grid of complex numbers:
z = x + j * y;

plot(real(z), imag(z), 'k .')
```

```
xlabel('Real Axis');
ylabel('Imaginary Axis');
title('Complex Number Plane');
grid on;

c = imag(1./z) > 1;

hold on;
plot(real(z(c)), imag(z(c)), 'r *')
axis([-1 1 -1.1 1])
title('Complex plane set Im(1/z) > 1')
```

When MCode-3.7 is run, we get the graph in Figure-3.17.

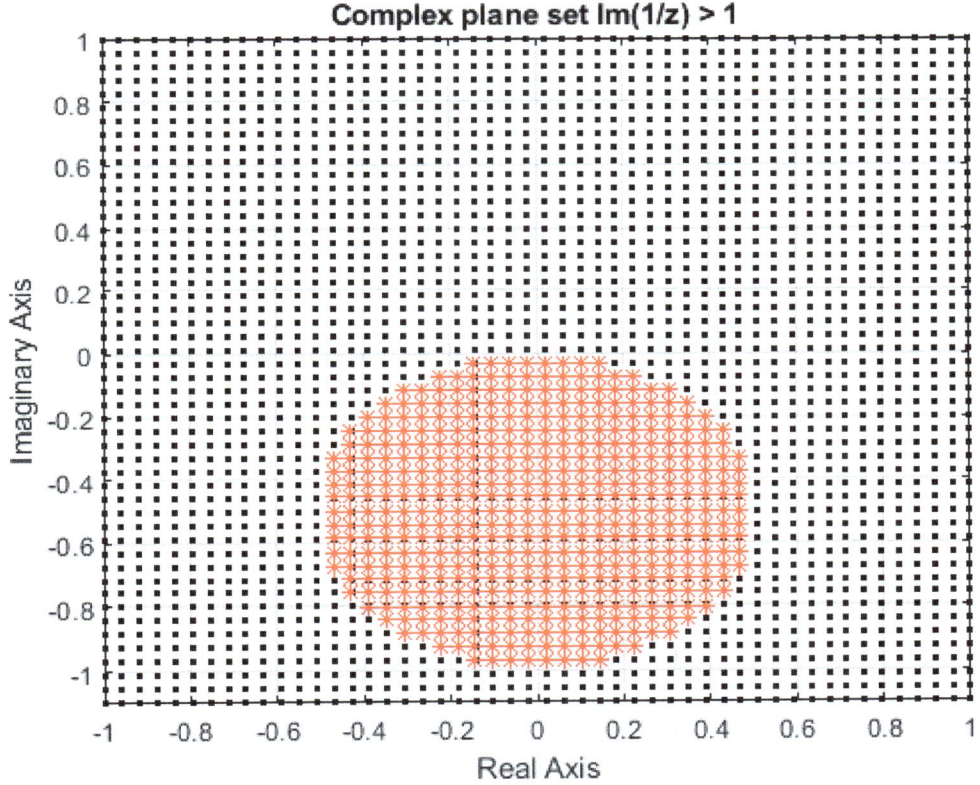

Figure-3.17

3.4 Line Integral in the Complex Plane

Contours

Conturs are paths over which complex integrations are taken. Consider the complex function

$$z(t) = x(t) + iy(t) \qquad a \leq t \leq b$$

where the real part $x(t)$ and imaginary part $y(t)$ are continuous functions of the real parameter t.

As t gets new values, the value of the complex number $z(t)$ changes.

Hiding t, the complex number can be written as

$$z = x + iy$$

Example-3.11: For the complex number set

$$z = \begin{cases} x + ix & \text{when} \quad 0 \leq x \leq 1 \\ x + i & \text{when} \quad 1 \leq x \leq 2 \end{cases}$$

the contour is depicted in Figure-3.18.
s

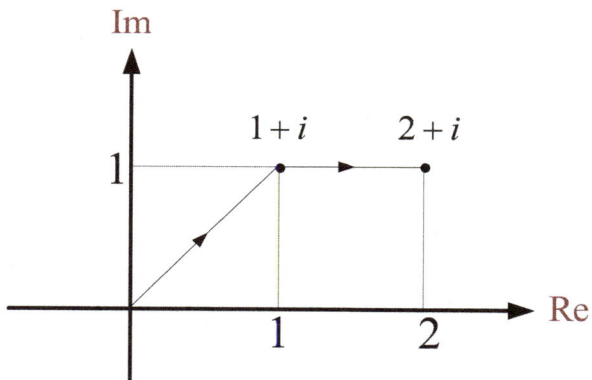

Figure-3.18

The contour can be drawn using the MATLAB program in MCode-3.8.

MCode-3.8

```
x1 = linspace(0, 1, 10);
x2 = linspace(1, 2, 10);
```

```
z = [x1+i*x1 x2+i];

plot(z);

axis([0 2.1 0 1.1]);
xlabel('x');
ylabel('y');
```

When MCode-3.8 is run, we get the graph in Figure-3.19.

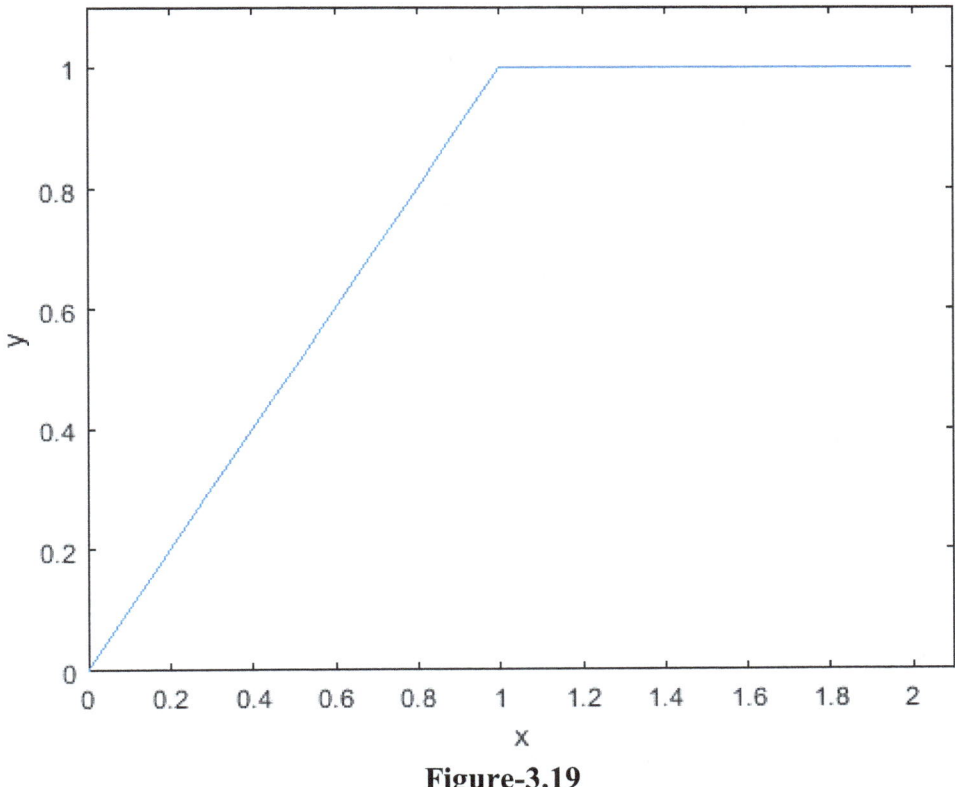

Figure-3.19

Example-3.12: For the complex number

$z(t) = x(t) + iy(t)$
we have
$$x = 2t, \quad y = t^2 \quad 0 \leq t \leq 2$$

The contour of the complex set is depicted in Figure-3.20.

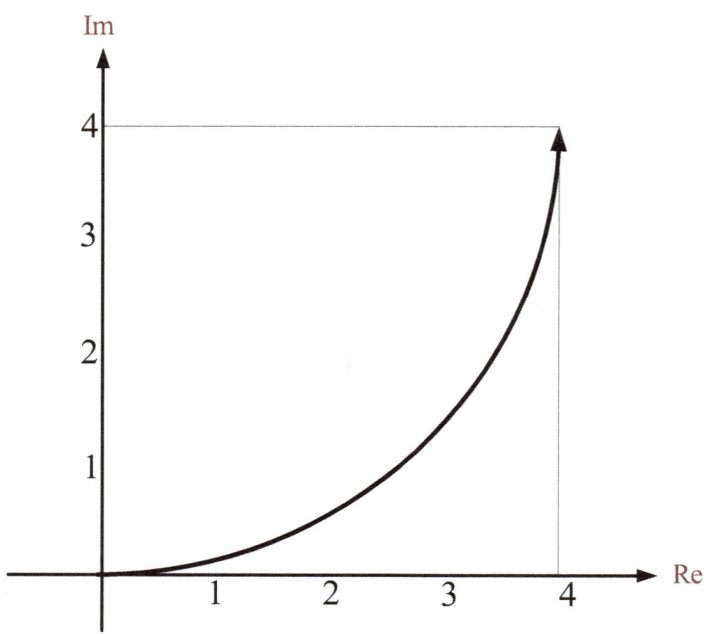

Figure-3.20

The contour can be drawn using the MATLAB program in MCode-3.9.

MCode-3.9

```
t = linspace(0, 2, 10);

x = 2*t;
y = t.^2;

z = x + i*y;

plot(z);

xlabel('x');
ylabel('y');
title('z = x+iy, x =2t, y=t^2, 0 \leq t \leq 2')
```

When MCode-3.9 is run, we get the graph in Figure-3.21.

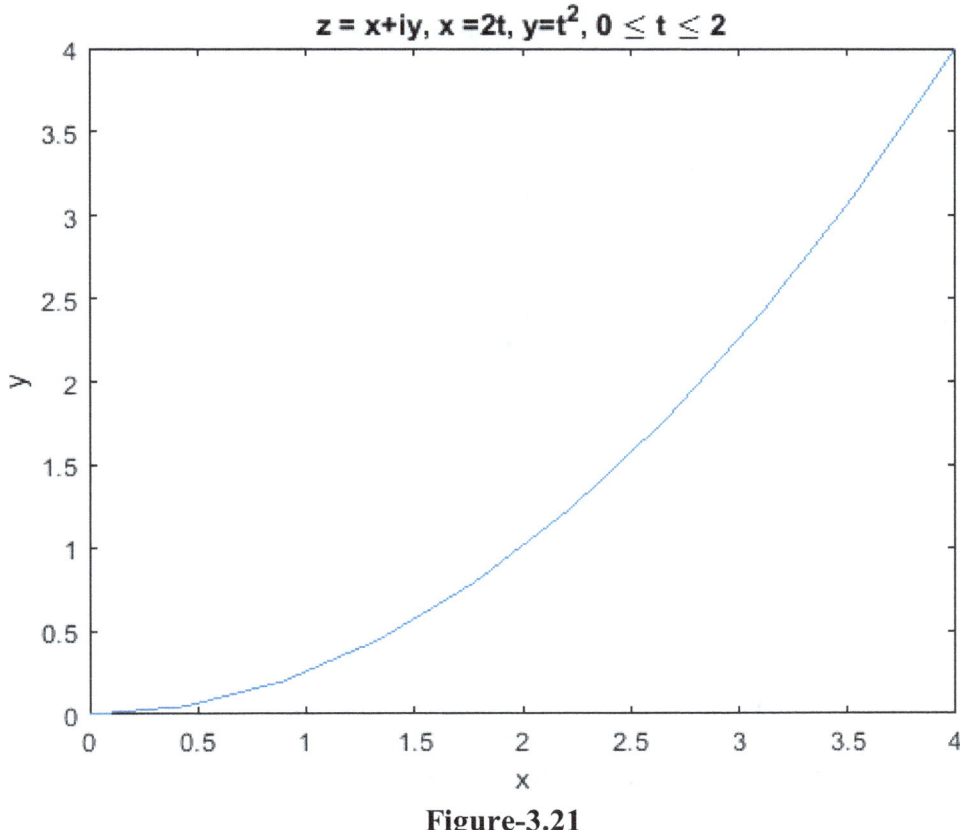

Figure-3.21

Example-3.13: The set of complex numbers expressed using

$$e^{it}, \qquad 0 \leq t < 3\pi$$

which is equal to

$$z(t) = \cos(t) + i\sin(t) \quad 0 \leq t < 3\pi$$

can be shown by the contour in Figure-3.22.

The curve is not closed and **the trace begins at 1 and ends at −1.**

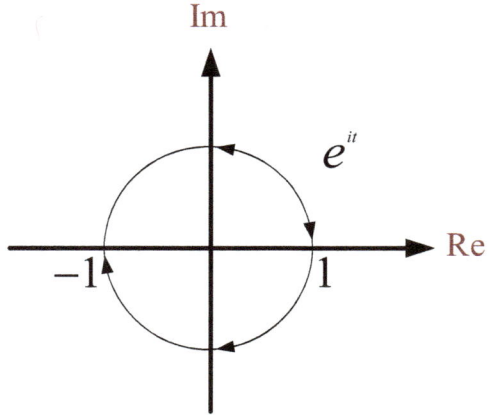

Figure-3.22

Note that the path of $z(t)$ is a contour and it can be denoted by the letter C.

The contour

$$z(t) = \cos(t) + i\sin(t) \quad 0 \leq t < 2\pi$$

can be drawn using the MATLAB code in MCode-3.10.

MCode-3.10

```
t = linspace(0, 2*pi, 1000);

x = cos(t);
y = sin(t);

z = x + i*y;

plot(z);

xlabel('x');
ylabel('y');
title('z = x+iy, x = cos(t), y = sin(t), 0 \leq t < 2\pi')
axis([-2 2 -2 2])
```

When MCode-3.10 is run, we get the graph in Figure-3.23.

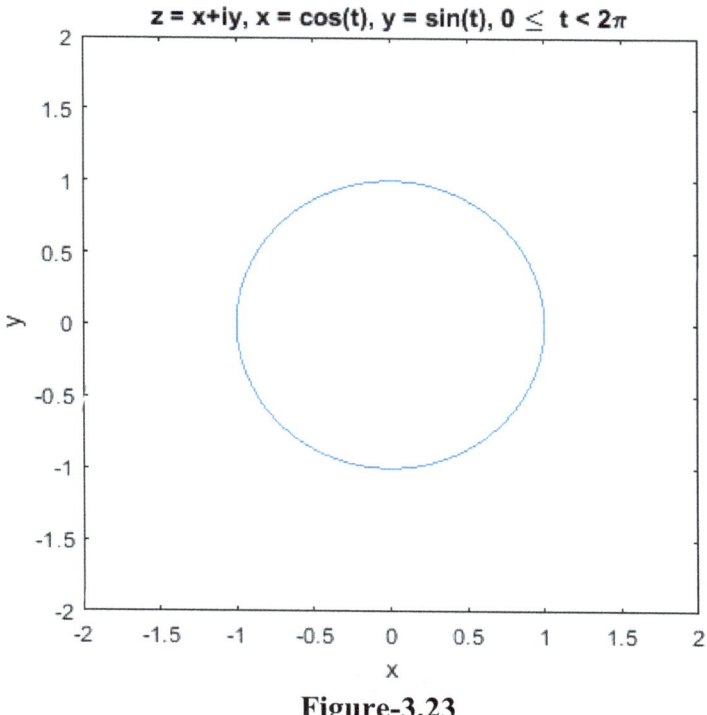

Figure-3.23

Example-3.14: Consider the contour shown in Figure-3.24. It is mathematically expressed as

$C_1 = e^{it}$ $\quad for \quad 0 \leq t \leq \pi$

$C_2 = -1 + it$ $\quad for \quad 0 \leq t \leq 2$

Implement this contour in MATLAB.

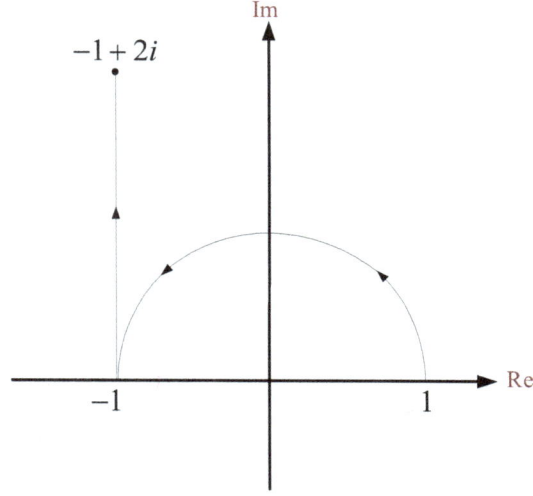

Figure-3.24

Solution-3.14: The contour can be implemented as in MCode-3.11.

MCode-3.11

```
t = linspace(0, pi, 1000);
c1 = exp(i*t); % C_1=e^it for 0≤t≤π

plot(c1);
hold on;

xlabel('Real');
ylabel('Imaginary');

t = linspace(0, 2, 1000);
c2 = -1 + i*t   % C_2=-1+it for  0≤t≤2
plot(c2);
axis([-1.1 1.1 0 2.1])
```

When MCode-3.11 is run, we get the graph in Figure-3.25.

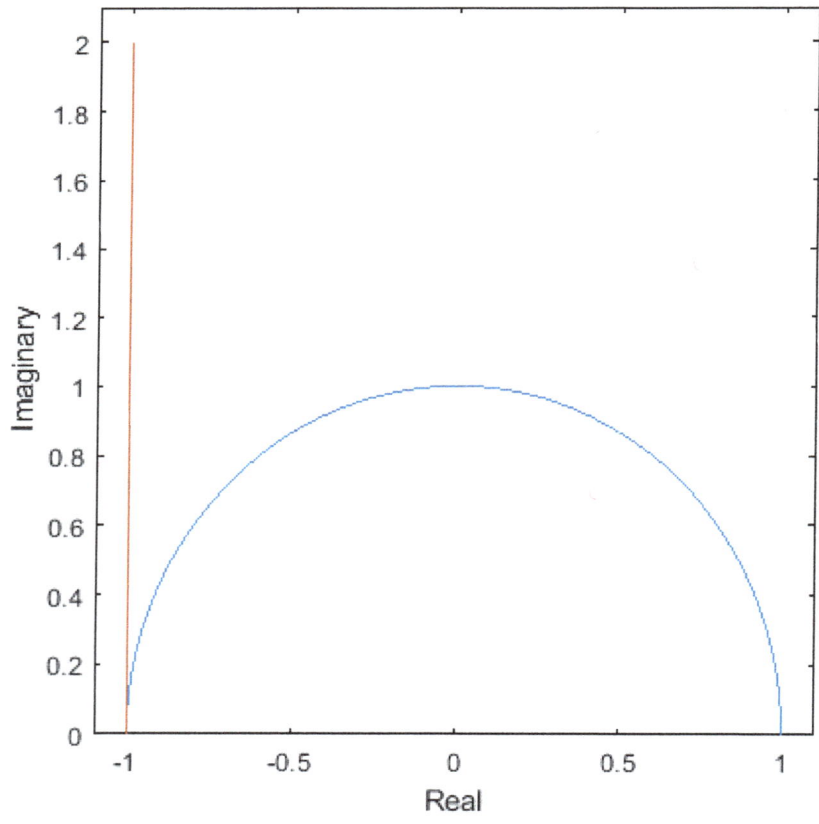

Figure-3.25

Example-3.15: A contour is mathematically expressed as

$$x(t) = 2\cos t \quad y(t) = 1 + 2\sin t \quad for \quad 0 \leq t < 2\pi$$
$$z(t) = x(t) + iy(t) \rightarrow z(t) = 2\cos t + i(1 + 2\sin t)$$

Implement this contour in MATLAB.

Solution-3.14: The contour can be implemented as in MCode-3.12.

MCode-3.12

```
t = linspace(0, 2*pi, 1000);
x_t = 2*cos(t);
y_t = 1+2*sin(t);
z_t = x_t +j*y_t;

plot(z_t);

xlabel('Real');
ylabel('Imaginary');
axis equal
```

When MCode-3.11 is run, we get the graph in Figure-3.26.

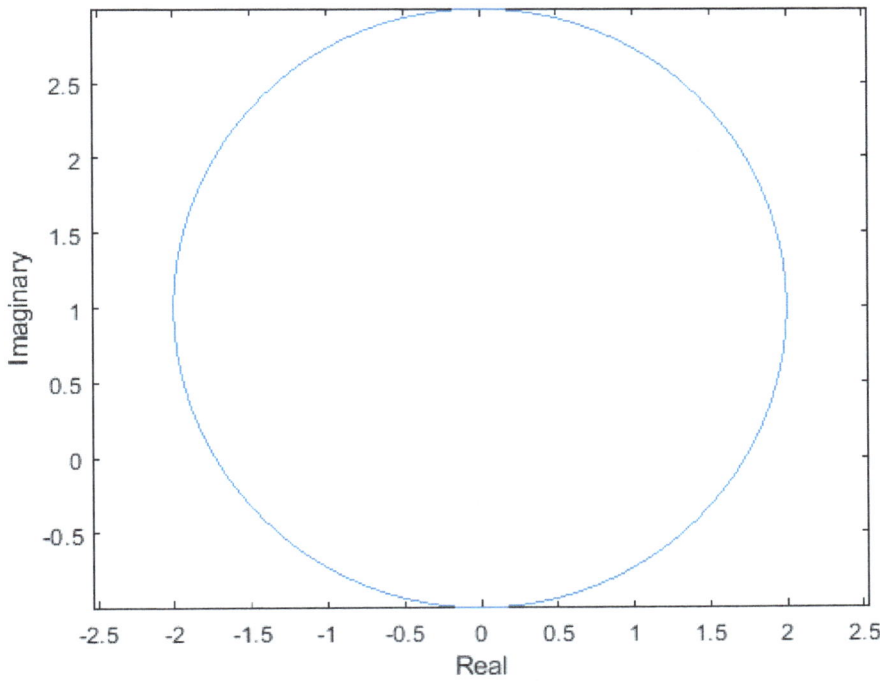

Figure-3.26

Contour Integration

Contour integrals, i.e., (complex) **line integrals,** over the path C is calculated as

$$\int_C f(z)dz \quad \text{or} \quad \int_{z_1}^{z_2} f(z)dz$$

where the integrand is integrated over a given curve C.

This curve C in the complex plane is called the path of integration.

The path C is represented by the

$$z(t) = x(t) + iy(t) \qquad a \leq t \leq b$$

Simple Connected

In a simply connected domain D, every simple closed curve encloses only points of D.

A circular disk is simply connected, on the other hand, an annulus is not simply connected.

3.4.1 Complex Integral Evaluation

First Evaluation Method: Indefinite Integration of Analytic Functions

The complex integral of $f(z)$ can be calculated as

$$\int_{z_0}^{z_1} f(z)dz = F(z_1) - F(z_0)$$

where
$$F'(z) = f(z)$$

Example-3.16: We can calculate

$$\int_0^{1+i} z^2 dz$$

as

$$\int_0^{1+i} z^2 dz = \frac{1}{3}z^3 \Big|_0^{1+i} \to \frac{1}{3}(1+i)^3 = -\frac{2}{3} + \frac{2}{3}i$$

Second Evaluation Method: Use of a Representation of a Path

Let C be the contour path, represented by $z(t)$, where $a \leq t \leq b$. Let $f(z)$ be a continuous function on C. Then, we have

$$\int_C f(z)dz = \int_a^b f[z(t)]\dot{z}(t)dt \qquad \dot{z} = \frac{dz}{dt}$$

For
$$\int_C f(z)dz$$
evaluation if

$$f(z) = u(t) + iv(t) \quad \text{and} \quad dz = dx + idy$$
then we can write

$$\int_C f(z)dz = \int_C [u(t) + iv(t)][dx + idy]$$

leading to

$$\int_C f(z)dz = \int_C [udx - vdy] + i\int_C [vdx + udy]$$

For the evaluation of
$$\int_a^b f[z(t)]\dot{z}(t)dt$$
If
$$f(z) = u(t) + iv(t) \quad \text{and} \quad z(t) = x(t) + iy(t)$$

then
$$\int_a^b f[z(t)]\dot{z}(t)dt = \int_C (u(t) + iv(t))(\dot{x}(t) + i\dot{y}(t))dt \rightarrow$$

$$\int_a^b f[z(t)]\dot{z}(t)dt = \int_C (udx - vdy) + i\int_C (vdx + udy)$$

Note: For the complex number i some books use j, that is $i^2 = 1$ or $j^2 = -1$

Example-3.17: Evaluate the integral

$$\int_C z^2 dz$$

along the path C shown in Figure-3.27.

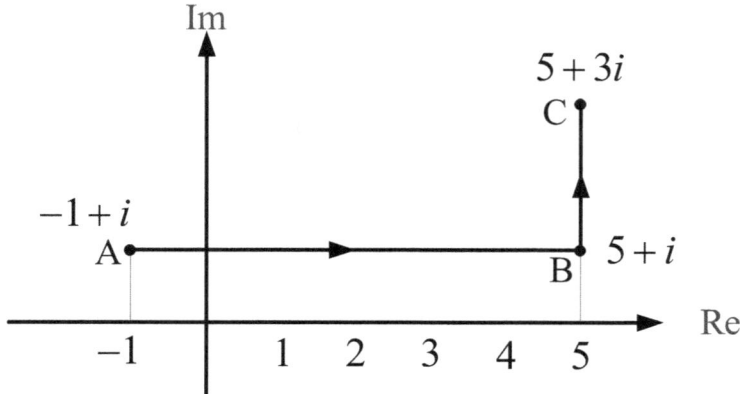

Figure-3.27

Solution-3.17: Let
$z = x + jy$

then we have
$f(z) = z^2 = (x + jy)^2 = (x^2 - y^2) + j2xy$
and

$dz = dx + jdy$

$f(z)dz = [(x^2 - y^2) + j2xy][dx + jdy] \to$
$f(z)dz = [(x^2 - y^2)dx - 2xydy] + j[2xydx + (x^2 - y^2)dy]$

Then,

$$\int_C f(z)dz = \int_C z^2 dz = \int_C [(x^2 - y^2)dx - 2xydy] + j\int_C [2xydx + (x^2 - y^2)dy]$$

Along AB $y = 1$ and $dy = 0$ so that

$$\int_{AB} z^2 dz = \int_{-1}^{5} (x^2 - 1)dx + j\int_{-1}^{5} 2xdx = 36 + 24j$$

Along BC $x = 5$ and $dx = 0$ so that

$$\int_{BD} z^2 dz = \int_{-1}^{3} -10 dy + j\int_{1}^{3} (25 - y^2) dy = -40 + \frac{124}{3}j$$

Thus

$$\int_C f(z)dz = \int_{AB} z^2 dz + \int_{BD} z^2 dz \rightarrow \int_C f(z)dz = -4 + \frac{196}{3}j$$

The evaluation of the integral can be performed as in MCode-3.13.

MCode-3.13

```
fun = @(z) z.^2;
C = [5+i];
F = integral(fun,-1+i,5+3*i,'Waypoints',C)  % -4.0000 +65.3333i
```

When MCode-3.13 is run, we get the result

$$-4.0000 + 65.3333i$$

which equals to the numerical evaluation result
$-4 + \frac{196}{3}j$

Example-3.18: Write the expression
$|z - 1 - 2j| = 2$

in rectangular form
$z = x + iy$

Solution-3.18: Using

$$|z - z_0| = r \quad \rightarrow \quad z = z_0 + re^{j\theta} \quad\quad 0 \leq \theta < 2\pi$$

for
$|z - 1 - 2j| = 2$
we get
$z = 1 + 2j + 2e^{j\theta}$
where using
$e^{j\theta} = \cos(\theta) + j\sin(\theta)$
we obtain

$z = 1 + 2j + 2(\cos(\theta) + j\sin(\theta))$
which can be rearranged as
$z = 1 + 2\cos(\theta) + j2(1 + \sin(\theta))$

Example-3.19: Evaluate the integral

$$\oint \frac{z+1}{z^3 - 2z^2} dz$$

around the circle expressed using
$|z - 1 - 2j| = 2$

where 2 is radius of the circle.

Solution-3.19: The expression
$|z - z_0| = r$
equals to
$z = z_0 + re^{j\theta} \qquad 0 \le \theta < 2\pi$

MATLAB program can be written as in MCode-3.14.

MCode-3.14

```
myFun=@(z)(z+1)./(z.^3-2*z.^2);
g=@(theta)(1+2*cos(theta))+1i*(2+2*sin(theta));
gp=@(theta)-2*sin(theta)+1i*2*cos(theta);
result=quad(@(t) myFun(g(t)).*gp(t),0,2*pi)
```

Example-3.20: Calculate the integral around the unit circle

$$\int_C \frac{dz}{z}$$

Solution-3.20: Let
$$f(z) = \frac{1}{z}$$
then we have
$$\int_C f(z)dz = \int_a^b f[z(t)]\,\dot{z}(t)dt$$

The unit circle can be expressed as

$$z(t) = \cos t + i\sin t = e^{it}$$

as shown in Figure-3.28, and we have
$$\dot{z}(t) = ie^{it}$$
and
$$f(z(t)) = \frac{1}{z(t)} = e^{-it}$$

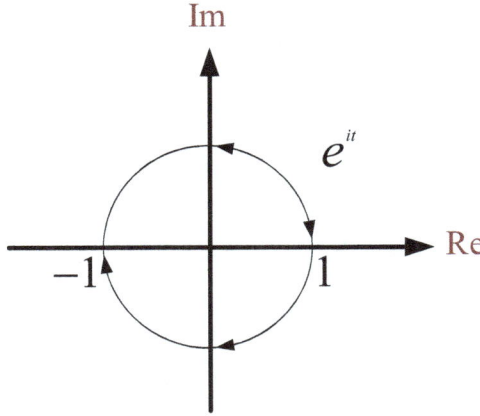

Figure-3.28

We can evaluate the integral as

$$\int_a^b f[z(t)]\,\dot{z}(t)dt = \int_0^{2\pi} e^{-it}ie^{it}dt = 2\pi i$$

Thus, we have
$$\int_C \frac{dz}{z} = 2\pi i$$
which can also be written as

$$\oint_C \frac{dz}{z} = 2\pi i$$

The integral calculation can be performed in MATLAB as in MCode-3.15.

MCode-3.15

```
myFunc=@(z)(1./z);
g=@(theta)(cos(theta)+j*sin(theta));
gp=@(theta)-sin(theta)+j*cos(theta);
```

```
result=quad(@(t)myFunc(g(t)).*gp(t),0,2*pi)  % 2*pi*i = 6.2832i

% or use, result=integral(@(t)myFunc(g(t)).*gp(t),0,2*pi) % 2*pi*i = 6.2832i
```

Example-3.21: Calculate the integral around the unit circle $|z - z_0| \leq r$

$$\oint_C (z - z_0)^m dz$$

where m is an integer and z_0 is a constant.

Solution-3.21: The integration path is shown in Figure-3.29.

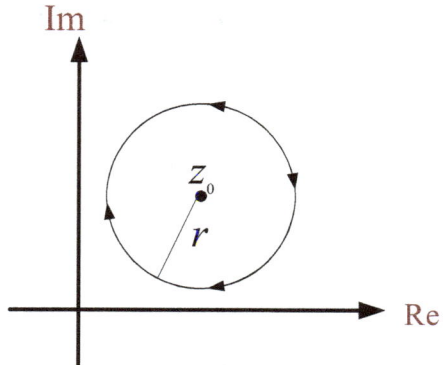

Figure-3.29

The circle can be expressed as

$$z(t) = z_0 + r(\cos t + i\sin t) = z_0 + re^{it} \quad \dot{z}(t) = rie^{it}$$

and
$$f(z) = (z - z_0)^m$$

We want to calculate
$$\oint_C f(z)dz$$

Using
$$\int_C f(z)dz = \int_a^b f[z(t)]\,\dot{z}(t)dt$$

We get

$$\int_a^b f[z(t)]\,\dot{z}(t)dt = \int_0^{2\pi} r^m e^{imt} r i e^{it} dt$$

$$= ir^{m+1} \int_0^{2\pi} e^{i(m+1)t} dt$$

where the expression

$$\int_0^{2\pi} e^{i(m+1)t} dt = \int_0^{2\pi} [\cos([m+1]t) + i\sin([m+1]t)]dt$$

equals 0 when $m \neq -1$, and for $m = -1$ we have

$$\int_0^{2\pi} e^{i(-1+1)t} dt = 2\pi$$

Then

$$ir^{m+1} \int_0^{2\pi} e^{i(m+1)t} dt = \begin{cases} 2\pi i & m = -1 \\ 0 & m \neq -1 \end{cases}$$

Thus,

$$\oint_C (z-z_0)^m dz = \begin{cases} 2\pi i & m = -1 \\ 0 & m \neq -1 \end{cases}$$

Example-3.22: Calculate the integral around the unit circle $|z - z_0| \leq 2$

$$\oint_C (z-z_0)^2 dz$$

where $z_0 = 1 + 2j$ using MATLAB.

Solution-3.22: We can write the MATLAB program as in MCode-3.16.

MCode-3.16

```
myFunc=@(z)(z-1-2*j).^2;
g=@(theta)(1+2*cos(theta))+1i*(2+2*sin(theta));
gp=@(theta)-2*sin(theta)+1i*2*cos(theta);
result=integral(@(t) myFunc(g(t)).*gp(t),0,2*pi)

%  0 -> -2.6645e-15 - 1.1102e-15i
```

Dependence on path

The result of a complex line integral may depend not only on the endpoints of the path but also on the path itself.

Example-3.23: Evaluate
$$\int_C Re(z)\, dz$$

along the paths C^* and $C_1 + C_2$ shown in Figure-3.30.

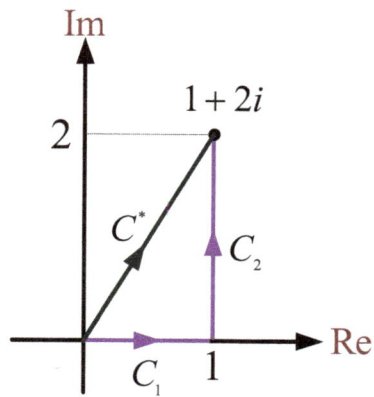

Figure-3.30

Solution-3.23: Let
$z = x + iy$
Then, we have

$$\int_C f(z)dz \to f(z) = Re(z) = x$$

We can write the mathematical expressions for the path C^* as

$C^*: x(t) = t \quad y(t) = 2it$

and we have
$z(t) = t + 2it$
and $\quad z'(t) = (1 + 2i)dt \quad\quad 0 \leq t < 1$

Similarly for the paths C_1 and C_2, we have

C_1: $\quad x(t) = t \quad\quad y(t) = 0 \quad z(t) = t \quad \dot{z}(t) = 1 \quad 0 \le t \le 1$

C_2: $\quad x(t) = 1 \quad\quad y(t) = t \quad z(t) = 1 + it \quad \dot{z}(t) = i \quad 0 \le t \le 2$

Then, using

$$\int_a^b f[z(t)]\,\dot{z}(t)\,dt$$

we get

$$C^*: \int_a^b f[z(t)]\,\dot{z}(t)\,dt = \int_0^1 t(1+2i)\,dt = \frac{1}{2} + i$$

and for the path $C_1 + C_2$, we get

$$C_1 + C_2: \int_a^b f[z(t)]\,\dot{z}(t)\,dt \rightarrow$$

$$\int_a^b f[z(t)]\,\dot{z}(t)\,dt = \int_{C_1} f[z(t)]\dot{z}(t)\,dt + \int_{C_2} f[z(t)]\dot{z}(t)\,dt$$

$$= \int_0^1 t\,dt + \int_1^2 i\,dt$$

$$= \frac{1}{2} + 2i$$

We can evaluate the integral using the MATLAB code MCode-3.17.

MCode-3.17

```
myFun = @(z)(real(z));
C = [1+2i]; % C*
F = integral(myFun,0,1+2i,'Waypoints',C) % 0.5000 + 1.0000i

%-----------------------------------------------------------

myFun = @(z)(real(z));
C = [1]; % C1+C2
F = integral(myFun,0,1+2i,'Waypoints',C) % 0.5000 + 2.0000i
```

Example-3.24: Let us evaluate the contour integral

$$\int_{C_1} \frac{dz}{z}$$

where C_1 is the top half
$z = e^{it} \quad 0 \le t \le \pi$

of the circle $|z| = 1$ from $z = 1$ to $z = -1$

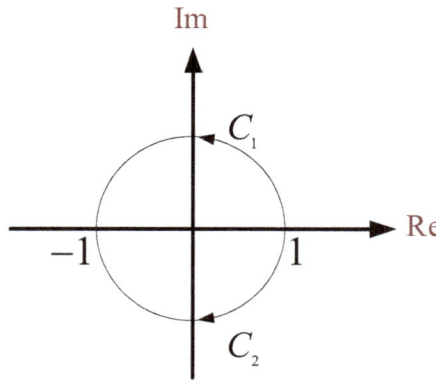

Figure-3.21

Solution-3.24:

$$\int_{C_1} \frac{dz}{z} = \int_0^{\pi} \frac{1}{e^{it}} i e^{it} dt = i \int_0^{\pi} dt = \pi i$$

We can evaluate the integral using the MATLAB code MCode-3.18.

MCode-3.18

```
myFunc=@(z)(1./z);
g=@(theta)(cos(theta)+i*sin(theta));
gp=@(theta)-sin(theta)+i*cos(theta);
result=integral(@(t) myFunc(g(t)).*gp(t),0,pi); %  3.1416i
```

Example-3.25: Let us evaluate the contour integral

$$\int_{C_2} \frac{dz}{z}$$

on C_2 which is shown in Figure-3.32.

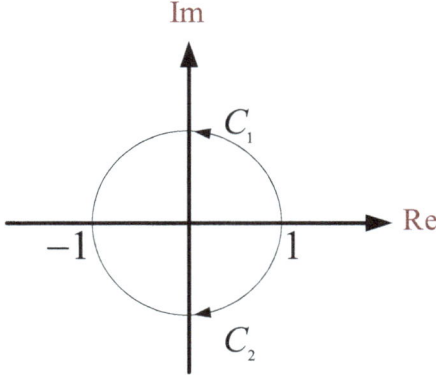

Figure-3.32

Solution-3.25: The path $-C_2$ can be represented as

$z = e^{it} \quad \pi \leq t \leq 2\pi$
Then,

$$\int_{C_2} \frac{dz}{z} = -\int_{-C_2} \frac{dz}{z} = -\int_\pi^{2\pi} \frac{1}{e^{it}} i e^{it} dt = -i \int_\pi^{2\pi} dt = -\pi i$$

If C is the closed curve $C = C_1 - C_2$, then

$$\int_C \frac{dz}{z} = \int_{C_1} \frac{dz}{z} - \int_{C_2} \frac{dz}{z} = \pi i - (-\pi i) = 2\pi i$$

Evaluation of
$$\int_{C_2} \frac{dz}{z}$$

can be done in MATLAB as in MCode-3.19.

MCode-3.19

```
myFunc=@(z)(1./z);
g=@(theta)(cos(theta)+i*sin(theta));
```

```
gp=@(theta)-sin(theta)+i*cos(theta);
result=integral(@(t) myFunc(g(t)).*gp(t),2*pi,pi)% -3.1416i, C2
```

Chapter-4

Logarithm, Power, Trigonometry of Complex Numbers, Cauchy Theorem

Abstract: In this chapter we explain how to calculate logarithm, power and trigonometric functions of complex numbers. Besides, Cauchy's theorems are explained in details with MATLAB applications.

4.1 Logarithm of Complex Numbers

The natural logarithm of the complex number

$z = x + iy$
is
$w = \ln z$
which equals
$e^w = z$
where using
$w = u + iv \quad z = re^{i\theta}$
we get
$e^{u+iv} = re^{i\theta}$
from which we obtain
$e^u = r \quad v = \theta$
leading to
$u = \ln r \quad v = \theta$
Then,
$w = \ln r + i\theta$

Thus, the natural logarithm of the complex number

$z = x + iy = re^{i\theta}$
equals
$w = \ln z \rightarrow w = \ln r + i\theta$

whicn can also be written as
$w = \ln z = \ln r + i\theta \pm i2n\pi$

where
$$\ln r + i\theta$$
can be denoted as

$$Ln\ z = \ln r + i\theta$$

which is called the principal value of $\ln z$. That is,

$$\ln z = Ln\ z \pm i2n\pi$$

Example-4.1: Calculate
$\ln(z)$
Solution-4.1:
$$z = re^\theta \rightarrow z = re^{(\theta \pm 2n\pi)} = re^\theta e^{\pm 2n\pi}$$

where θ is the principal argument and we have

$$\ln(z) = Ln\ z \pm 2n\pi$$
Example-4.2: Given
$z = 3 - 4i$
find $\ln z$

Solution-4.2: Let's first write the polar form of z as

$$z = |z|e^{i\theta} \rightarrow z = \sqrt{3^2 + 4^2}e^{i\theta} \rightarrow z = 5e^{i\theta} \quad \theta = -\arctan\frac{4}{3}$$

Then, we can evaluate principal value of $\ln z$ as

$$Ln\ z = \ln 5 + i\left(-\arctan\frac{4}{3}\right) \rightarrow Ln\ z = 1.609438 - 0.927295i$$

and $\ln z$ can be written as

$$\ln z = Ln\ z \pm 2n\pi i$$
leading to

$$\ln z = \ln 5 + i\left(-\arctan\frac{4}{3}\right) \pm 2n\pi i$$

$$\ln z = 1.609438 - 0.927295i \pm 2n\pi i$$

We can evaluate the Ln of a complex number using the MATLAB code MCode-4.1.

MCode-4.1

```
z = 3-4i;

w = log(z)  % 1.6094 - 0.9273i
```

4.2 General Powers

$\ln z = \text{Ln } z \pm 2n\pi i$

$z = e^{\ln z}$

General power of the complex number $z = x + iy$ is defined as

$z^c = e^{c \ln z}$

Since $\ln z$ is infinitely many-valued, z^c will, in general, be multivalued. The particular value

$z^c = e^{c \, \text{Ln } z}$

is called the principal value of z^c

Example-4.3: Calculate i^i

Solution-4.3:
$i^i = e^{i \ln i} = \exp(i \ln i)$
where

$i = re^{(\theta \pm 2n\pi)i} \rightarrow i = e^{\left(\frac{\pi}{2} \pm 2n\pi\right)i}$

$\ln i = \left(\frac{\pi}{2} \pm 2n\pi\right)i$

Thus,
$i^i = e^{i \ln i} = \exp\left(i\left(\frac{\pi}{2} \pm 2n\pi\right)i\right)$

$= \exp\left(-\left(\frac{\pi}{2} \pm 2n\pi\right)\right)$

$= e^{-\frac{\pi}{2} \mp 2n\pi}$

That is;

$$i^i = e^{-\frac{\pi}{2} \pm 2n\pi}$$

If you consider only principal value, the

$$i^i = e^{-\frac{\pi}{2}}$$

We can evaluate
i^i

number using the MATLAB code MCode-4.2

MCode-4.2

```
z = i^i; % 0.7214

exp(-pi/2)-z % = 0
```

Example-4.4: Calculate
$(1+i)^{2-i}$

Solution-4.4:
$$(1+i)^{2-i} = e^{(2-i)\ln(1+i)}$$
$$= \exp((2-i)\ln(1+i))$$

where using

$$1+i = \sqrt{2}\, e^{i\left(\frac{\pi}{4} \pm 2n\pi\right)}$$

we obtain
$$(1+i)^{2-i} = e^{(2-i)\ln(1+i)}$$
$$= \exp((2-i)\ln(1+i))$$
$$= \exp\left((2-i)\ln\left(\sqrt{2}\, e^{i\left(\frac{\pi}{4} \pm 2n\pi\right)}\right)\right)$$
$$= \exp\left((2-i)\left[\ln\sqrt{2} + \frac{\pi}{4}i \pm 2npi\right]\right)$$

We can evaluate
$(1+i)^{2-i}$

number using the MATLAB code MCode-4.3.

```
z =(1+i)^(2-i)  %  1.4900 + 4.1257i
```

Example-4.5: For the function
$$f(z) = z^{-1+i}$$
calculate
$$\int_C f(z)dz$$

where C is the unit circle path,

$$z = e^{it} \quad -\pi \le t < \pi$$

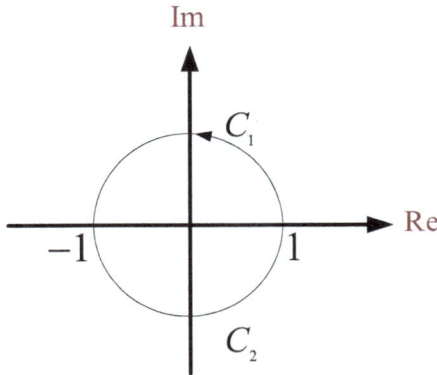

Figure-4.1

Solution-4.5:
$$z = e^{it} \quad -\pi \le t < \pi$$

$$\int_C f(z)dz = \int_C f(z(t))\dot{z}(t)dt$$

$$f(z) = z^{-1+i} = \exp([-1+i]\ln z)$$
where using
$$z = e^{it}$$

$$f(z(t)) = z^{-1+i} = \exp([-1+i](it))$$

Then

$$f(z(t))\dot{z}(t) = \exp([-1+i](it))\, i\exp(it)$$
$$= i\,e^{-t}$$

Thus,

$$\int_C f(z(t))\dot{z}(t)dt = \int_{-\pi}^{\pi} i\,e^{-t}dt = i(-e^{-\pi} + e^{\pi}) = i2\sin h\pi$$

We can evaluate the integral using the MATLAB code MCode-4.4.

MCode-4.4

```
myFunc=@(z)(z.^(-1+i));
g=@(theta)(cos(theta)+i*sin(theta));
gp=@(theta)-sin(theta)+i*cos(theta);
result=integral(@(t) myFunc(g(t)).*gp(t),-pi,pi) % 23.0975i

i*2*sinh(pi)-result % 0
```

4.3 Trigonometric Complex Functions

The well known trigonometric complex functions can be written as

$$\cos z = \frac{1}{2}(e^{iz} + e^{-iz})$$

$$\sin z = \frac{1}{2i}(e^{iz} - e^{-iz})$$

$$\tan z = \frac{\sin z}{\cos z} \qquad \cot z = \frac{\cos z}{\sin z}$$

$$\sec z = \frac{1}{\cos z} \qquad \csc z = \frac{1}{\sin z}$$

and we have the derivative properties

$$(\cos z)' = -\sin z \qquad (\sin z)' = \cos z \qquad (\tan z)' = \sec^2 z$$

Example-4.6: Calculate $\sin(1+i)$

Solution-4.6: The sine function $\sin(1+i)$ can be written as

$$\sin(1+i) = \frac{1}{2i}\left(e^{i(1+i)} - e^{-i(1+i)}\right)$$

where using

$$e^{i(1+i)} = e^i e^{-1} \rightarrow e^{i(1+i)} = [\cos(1) + i\sin(1)]e^{-1}$$

$$e^{-i(1+i)} = e^{-i}e^{1} \rightarrow e^{-i(1+i)} = [\cos(-1) + i\sin(-1)]e^{1}$$

we get

$$-0.5i\left([\cos(1) + i\sin(1)]e^{-1} - [\cos(-1) + i\sin(-1)]e^{1}\right)$$

which can be written as

$$0.5\left[(\sin(1)e^{-1} - \sin(-1)e^{1}) + i(\cos(-1)e^{1} - \cos(1)e^{-1})\right]$$

We can calculate $\sin(1+i)$ using the MATLAB code MCode-4.5.

MCode-4.5

```
z = 1 + i;

r1 = sin(1)*exp(-1)-sin(-1)*exp(1);
r2 = j*(cos(-1)*exp(1)-cos(1)*exp(-1));
r = 0.5*(r1+r2);
sin(z)  % 1.2985 + 0.6350i

r-sin(z)  % 0
```

4.3.1 Hyperbolic Trigonometric Complex Functions

The hyperbolic trigonometric complex functions can be written as

$$\sinh z = \frac{1}{2}(e^z - e^{-z}) \qquad \cosh z = \frac{1}{2}(e^z + e^{-z})$$

$$\tanh z = \frac{\sinh z}{\cosh z} \qquad \coth z = \frac{\cosh z}{\sinh z}$$

$$\tanh z = \frac{\sinh z}{\cosh z} \qquad \coth z = \frac{\cosh z}{\sinh z}$$

$$\text{sech } z = \frac{1}{\cosh z} \qquad \text{csch } z = \frac{1}{\sinh z}$$

and we have the derivative properties

$$(\sinh z)' = \cosh z \qquad (\cosh z)' = \sinh z$$

The complex trigonometric functions are related to the complex hyperbolic functions as

$$\sinh iz = i\sin z \qquad \cosh iz = \cos z$$
and
$$\sin iz = i\sinh z \qquad \cos iz = \cosh z$$

4.4 Bounds for Complex Integrals

The absolute of a complex integral can be upper bounded as

$$\left| \int_C f(z)dz \right| \leq KL$$

where L is the length of C, and K is a constant such that $|f(z)| \leq K$ on C.

Example-4.7: Find an upper bound for the absolute value of the integral

$$\int_C z^2 dz$$

where C is the path shown below

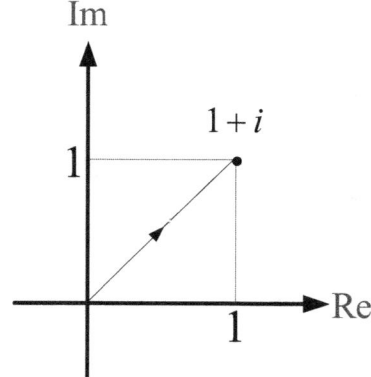

Figure-4.2

Solution-4.7:
$L = Path\ length = \sqrt{2}$

$|f(z)| = |z^2| \leq 2$ on the path C

Then

$$\left| \int_C z^2 dz \right| \leq 2\sqrt{2} = 2.8284$$

In MCode-4.6, we illustrate the use of the bound expression.

MCode-4.6

```
myFun = @(z)(z.^2);
C = [1+i];
F = integral(myFun,0,1+i,'Waypoints',C); % -0.6667 + 0.6667i

pVec = [0 0.25+0.25i 0.5+0.5i 0.75+0.75i 1+1i];

myFunVals = myFun(pVec);

mVal = max(abs(myFunVals));

abs(F) < mVal*2 % result is true
```

4.5 Cauchy's Integral Theorem

In this section we explain Cauchy's integral theorem. However, before explaining the theorem we provide some fundamental definitions.

4.5.1 Simple Closed Path

A simple closed path does not intersect itself as shown in

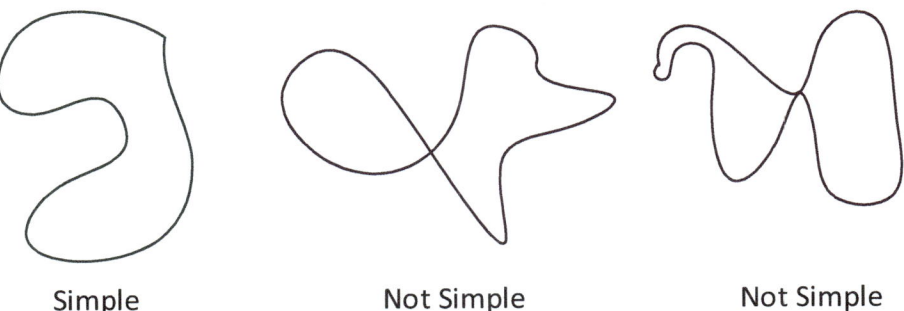

Simple   Not Simple   Not Simple

Figure-4.3

4.5.2 Simply Connected Domain

If every closed path in a domain D encloses only points in then the domain D is called a simple connected domain. Multiply connected domain is a domain that is not simply connected.

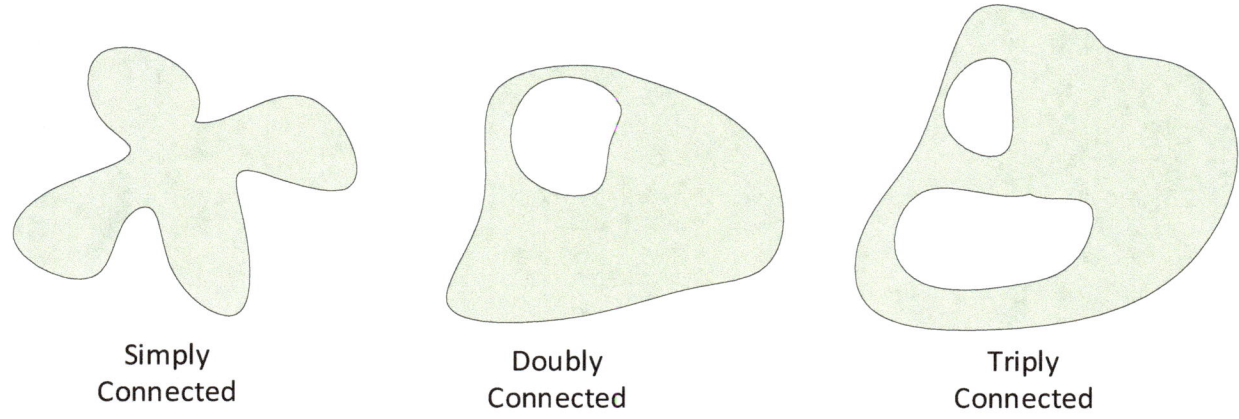

Simply Connected   Doubly Connected   Triply Connected

Figure-4.4

A function $f(z)$ is said to be **analytic** in a domain D, if $f(z)$ is differentiable at all points of D.

The function $f(z)$ is analytic at a point $= z_0$, if $f(z)$ is differentiable at z_0.

4.5.3 Cauchy's Integral Theorem

If $f(z)$ is analytic in a simply connected domain D, then for any closed path C in D as shown in Figure-4.5, we have

$$\oint_C f(z) dz = 0$$

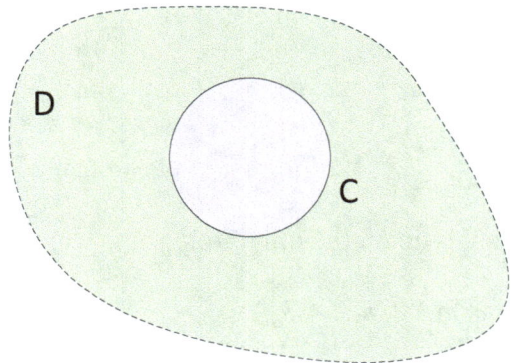

Figure-4.5

Example-4.8: Consider the entire Z-plane, for this plane we have

$$\oint_C e^z dz = 0 \qquad \oint_C \cos z\, dz = 0 \qquad \oint_C z^n dz = 0 \qquad n = 0, 1, \ldots$$

These integrals numerically can be evaluated in MATLAB as in MCode-4.7.

MCode-4.7

```
myFunc=@(z)(exp(z));
g=@(theta)(cos(theta)+i*sin(theta));
gp=@(theta)-sin(theta)+i*cos(theta);
result=integral(@(t) myFunc(g(t)).*gp(t),0, 2*pi) % result is 0
%-----------------------------------------------------------------

myFunc=@(z)(cos(z));
g=@(theta)(cos(theta)+i*sin(theta));
gp=@(theta)-sin(theta)+i*cos(theta);
result=integral(@(t) myFunc(g(t)).*gp(t),0, 2*pi) % result is 0
%-----------------------------------------------------------------

myFunc=@(z)(z.^3);
g=@(theta)(cos(theta)+i*sin(theta));
gp=@(theta)-sin(theta)+i*cos(theta);
result=integral(@(t) myFunc(g(t)).*gp(t),0, 2*pi) % result is 0
```

Example-4.8: The integral

$$\oint_C \bar{z}\, dz$$

can be calculated around the unit circle C as

$$\oint_C \bar{z}\,dz = \int_0^{2\pi} e^{-it} i e^{it} dt = 2\pi i$$

The function $f(z) = \bar{z}$ is not analytic on the unit circle.

The integral can be evaluated as in MCode-4.8.

MCode-4.8

```
myFunc=@(z)(conj(z));
g=@(theta)(cos(theta)+i*sin(theta));
gp=@(theta)-sin(theta)+i*cos(theta);
result=integral(@(t)myFunc(g(t)).*gp(t),0, 2*pi) % result is 6.2832i
```

Example-4.9: Analyticity is a sufficient but not necessary condition for the integral of a function to be zero. For instance, we have

$$\oint_C \frac{dz}{z^2} = 0$$

where C is the unit circle.

However, $f(z) = 1/z^2$ is not analytic at $z = 0$. We do not use Cauchy's theorem in this example.

Hence, analyticity of f in D is sufficient for

$$\oint_C f(z)\,dz = 0$$

to be true but it is not necessary condition. The integral can be evaluated as in MCode-4.9.

MCode-4.9

```
myFunc=@(z)(z.^-2);
g=@(theta)(cos(theta)+i*sin(theta));
gp=@(theta)-sin(theta)+i*cos(theta);
result=integral(@(t)myFunc(g(t)).*gp(t),0, 2*pi) % result is 0
```

Example-4.10: We have

$$\oint_C \frac{dz}{z} = 2\pi i$$

The function $1/z$ is not analytic at point $z=0$.

The integral can be evaluated as in MCode-4.10.

MCode-4.10

```
myFunc=@(z)(1./z);
g=@(theta)(cos(theta)+1*sin(theta));
gp=@(theta)-sin(theta)+i*cos(theta);
result=integral(@(t) myFunc(g(t)).*gp(t),0, 2*pi) % result is 6.2832i
```

4.5.4 Independence of Path

Theorem

If $f(z)$ is analytic in a simply connected domain D, then the integral of $f(z)$ gives the same result for different paths if the end points of the paths are the same.

Proof: Considering Figure-ZZ, according to Cauchy's integral theorem, we can write

$$\int_{C_1} f(z)dz + \int_{C_2^*} f(z)dz = 0 \qquad (4.1)$$

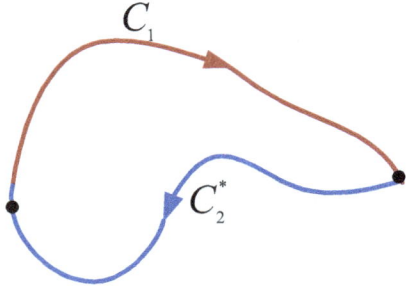

Figure-4.6

The path C_2^* in Figure-1 can be drawn as in Figure-2 and we have $C_2 = -C_2^*$

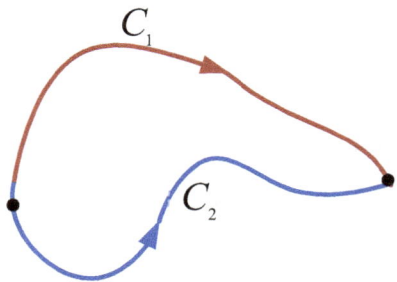

Figure- 4.7

Using (4.1) we can write

$$\int_{C_1} f(z)dz + \int_{-C_2} f(z)dz = 0$$

which can be rearranged as

$$\int_{C_1} f(z)dz - \int_{C_2} f(z)dz = 0$$

leading to

$$\int_{C_1} f(z)dz = \int_{C_2} f(z)dz$$

Thus in general, if $f(z)$ is analytic on path C_1 and C_2 and paths have the same beginning and end points, then we can write

$$\int_{C_1} f(z)dz = \int_{C_2} f(z)dz$$

In MCode-4.11, we calculate

$$\int_C e^z dz$$

on the paths $0 \leq z \leq \pi$ and $2\pi \leq z \leq -\pi$ and show that the results are the same.

MCode-4.11

```
myFunc=@(z)(exp(z));
g=@(theta)2*(cos(theta)+i*sin(theta));
gp=@(theta)2*(-sin(theta)+i*cos(theta));

result=integral(@(t)myFunc(g(t)).*gp(t),0,pi)     %  -7.2537

result=integral(@(t)myFunc(g(t)).*gp(t),2*pi,-pi) %  -7.2537
```

4.6 Cauchy's Integral Theorem for Multiply Connected Domains

Cauchy's theorem applies to multiply connected domains.

If $f(z)$ is analytic in any domain D^* that contains D and its boundary curves as shown in Figure-4, we claim that

$$\oint_{C_1} f(z)dz = \oint_{C_2} f(z)dz$$

both integrals being taken counterclockwise (or both clockwise)

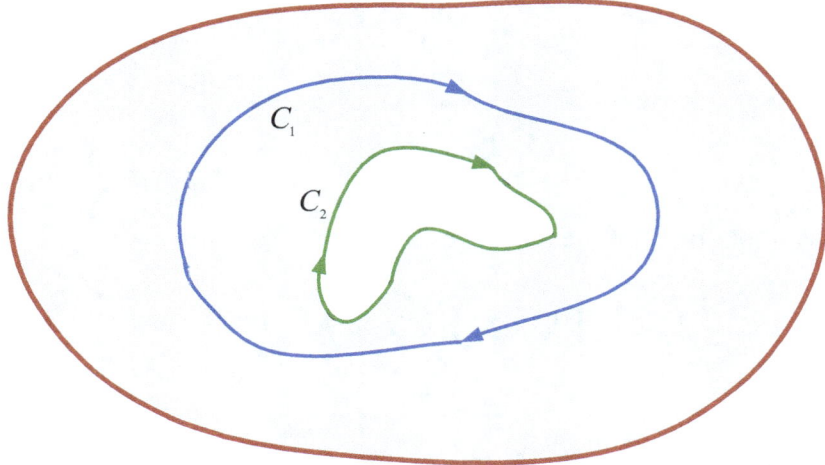

Figure-4.8

Proof: We can label the path segments as shown in Figure-4.9.

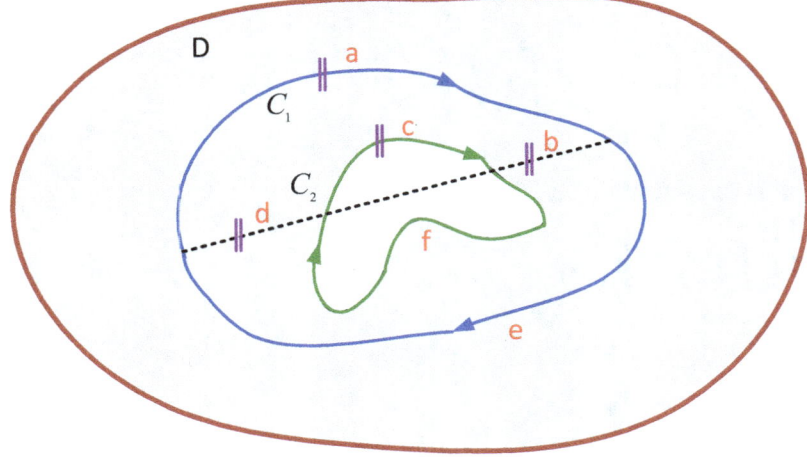

Figure-4.9

We can divide the paths into two parts and show only the upper segments as in Figure-4.10.

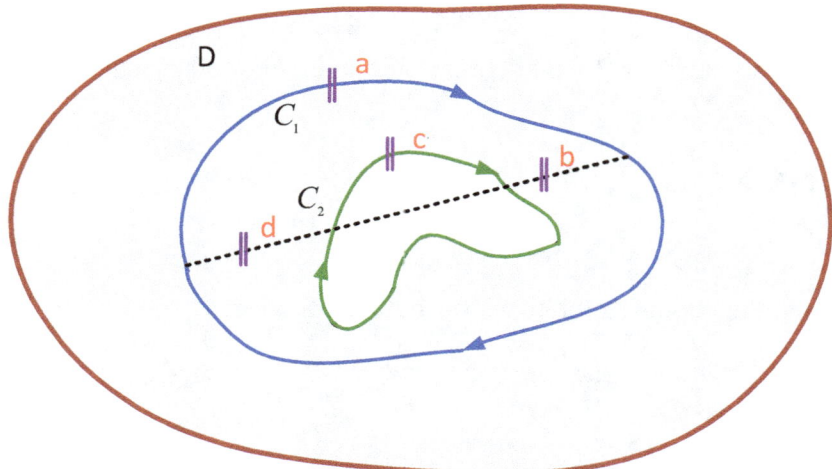

Figure-4.10

In Figure-5, for the upper parts of the paths we can write

$$\int_a (\cdot) + \int_b (\cdot) + \int_{-c} (\cdot) + \int_d (\cdot) = 0 \qquad (4.2)$$

In a similar manner for the lower paths shown in Figure-4.11 we can write

$$\int_{-f} (\cdot) + \int_{-b} (\cdot) + \int_e (\cdot) + \int_{-d} (\cdot) = 0 \qquad (4.3)$$

Since

$$\int_{-b} (\cdot) = -\int_b (\cdot) \qquad \int_{-d} (\cdot) = -\int_d (\cdot)$$

When (4.21) and (4.3) are summed, we get

$$\int_a (\cdot) + \int_e (\cdot) = -\int_c (\cdot) - \int_f (\cdot)$$

which equals

$$\oint_{C_1} (\cdot) = \oint_{C_2} (\cdot)$$

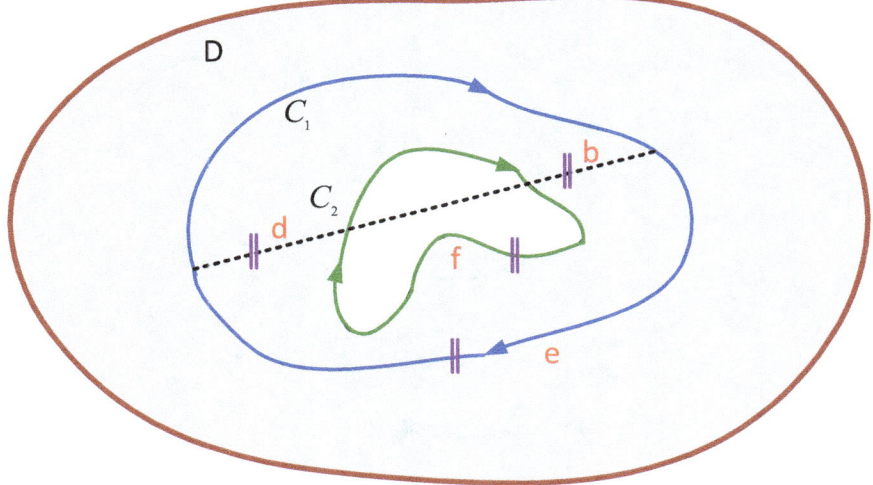

Figure-4.11

Theorem: Let C be a simple closed contour with counterclockwise direction, and C_k, $k=1,2..,n$ are simple disjoint closed contours with counterclockwise direction and all these contours are interior to the C. If f is analytic on all of these contours, and on the domain D, then we have

$$\int_C f(z)dz = \sum_{k=1}^{n} \int_{C_k} f(z)dz$$

4.6.1 Cauchy's Integral Formula

Theorem

$f(z)$ is an analytic function in a simply connected domain D. Then for any z_0 point of D, and for any simple closed path C which encloses z_0 as shown in Figure-4.12, we have

$$\oint_C \frac{f(z)}{z-z_0} dz = -2\pi i f(z_0)$$

which can also be written as

$$f(z_0) = -\frac{1}{2\pi i} \oint_C \frac{f(z)}{z-z_0} dz$$

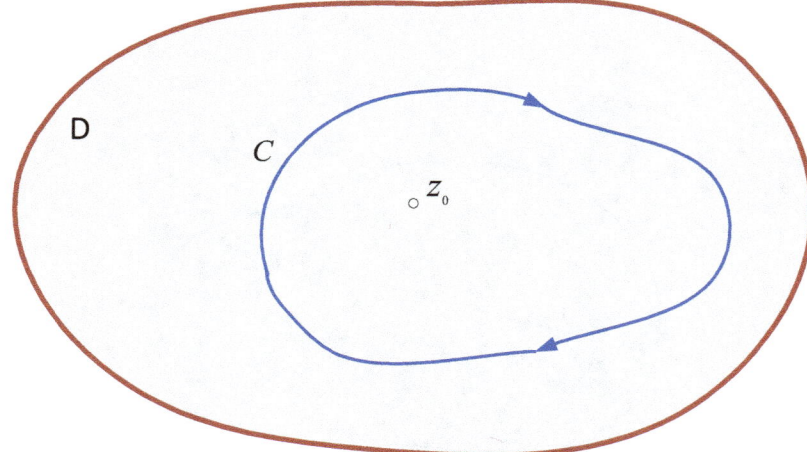

Figure-4.12

For counterclockwise direction as shown in Figure-4.13, we have

$$\oint_C \frac{f(z)}{z-z_0} dz = +2\pi i f(z_0)$$

which can also be written as

$$f(z_0) = +\frac{1}{2\pi i} \oint_C \frac{f(z)}{z-z_0} dz$$

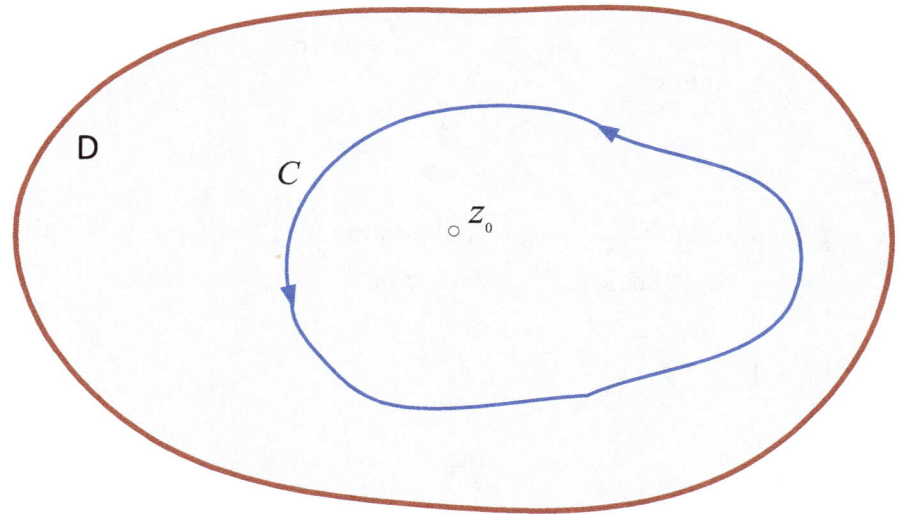

Figure-4.13

Example-4.11: Calculate

$$\oint_C \frac{e^z}{z-2} dz$$

along counterclockwise C which has an interior point at $z = 2$.

Solution-4.11:

Remember, previously we have found that

$$\oint_C (z-z_0)^m dz = \begin{cases} 2\pi i & m = -1 \\ 0 & m \neq -1 \end{cases}$$

That is,

$$\oint_C \frac{1}{z-2} dz = 2\pi i$$

Now using,

$$\oint_C \frac{f(z)}{z-z_0} dz = 2\pi i f(z_0)$$

for

$$\oint_C \frac{e^z}{z-2} dz$$

we get

$$\oint_C \frac{e^z}{z-2} dz = 2\pi i e^2$$

The integral can be evaluated in MATLAB as in MCode-4.12.

MCode-4.12

```
myFunc=@(z)(exp(z)./(z-2));
g=@(theta)4*(cos(theta)+i*sin(theta));
gp=@(theta)4*(-sin(theta)+i*cos(theta));

result=integral(@(t)myFunc(g(t)).*gp(t),0, 2*pi)  %  46.4268i
```

```
2*i*pi*(exp(1)^2)  %  46.4268i
```

Example-4.12: Calculate
$$\oint_C \frac{z^3 - 6}{2z - i} dz$$
The complex number
$$z_0 = \frac{1}{2}i$$
is inside C

Solution-4.12: The expression
$$\oint_C \frac{z^3 - 6}{2z - i} dz$$
can be written as
$$\oint_C \frac{\frac{1}{2}z^3 - 3}{z - \frac{1}{2}i} dz$$
and using
$$\oint_C \frac{f(z)}{z - z_0} dz = 2\pi i f(z_0)$$
we get
$$\oint_C \frac{\frac{1}{2}z^3 - 3}{z - \frac{1}{2}i} dz = 2\pi i \left(\frac{1}{2}z^3 - 3\right)\bigg|_{z=\frac{i}{2}}$$
$$= \frac{\pi}{8} - 6\pi i$$

We can calculate the given integral using the MATLAB code in MCode-4.13.

MCode-4.13

```
myFunc=@(z)((0.5*z.^3-3)./(z-0.5i));
g=@(theta)(cos(theta)+i*sin(theta));
gp=@(theta)(-sin(theta)+i*cos(theta));

result=integral(@(t)myFunc(g(t)).*gp(t),0, 2*pi)  %  0.3927 -18.8496i

pi/8-6*pi*i  %  0.3927 -18.8496i
```

Example-4.13: Write mathematical expressions for each of the circles in Figure-4.14.

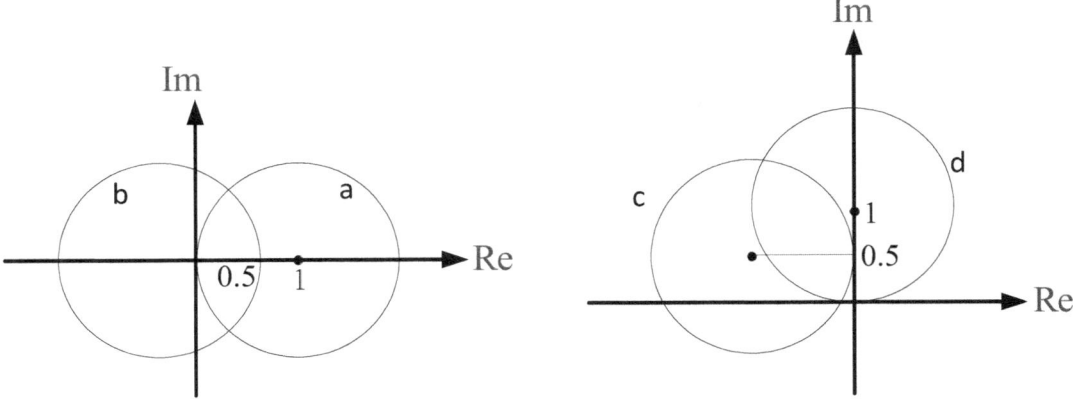

Figure-4.14

Solution-4.13: Using
$$|z - z_0| = r \rightarrow z = z_0 + re^{j\theta} \quad 0 \leq \theta < 2\pi$$

We can express the circles as

Circle a:

$|z - 1| = 1 \rightarrow z = 1 + e^{j\theta} \quad 0 \leq \theta < 2\pi$
Circle b:

$|z - 0.5| = 1 \rightarrow z = 0.5 + e^{j\theta} \quad 0 \leq \theta < 2\pi$

Circle c:

$|z + 1 - 0.5j| = 1 \rightarrow z = -1 + 0.5j + e^{j\theta} \quad 0 \leq \theta < 2\pi$
Circle d:

$|z - 0.5j| = 1 \rightarrow z = 0.5j + e^{j\theta} \quad 0 \leq \theta < 2\pi$

Example-4.14: Calculate the integral of

$$g(z) = \frac{z^2 + 1}{z^2 - 1} = \frac{z^2 + 1}{(z + 1)(z - 1)}$$

counterclockwise around each of the four circles in Figure-4.15.

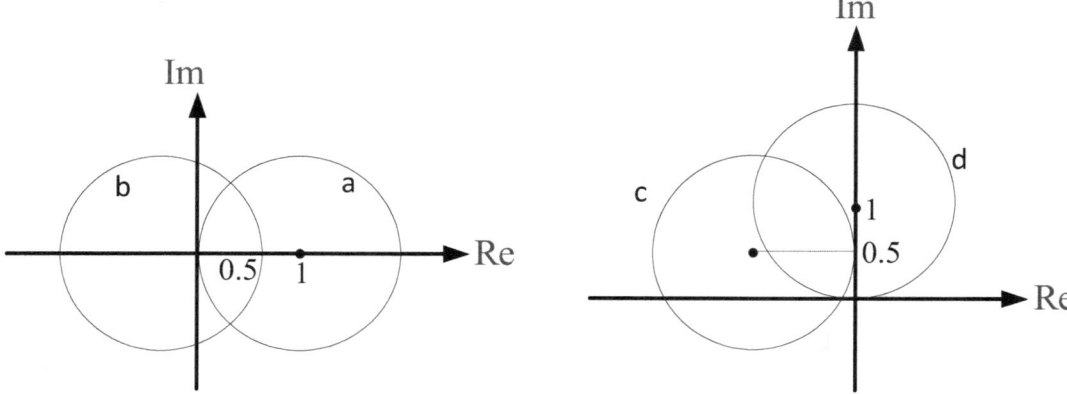

Figure-4.15

Solution-4.14: $g(z)$ is not analytic at -1 and 1. These are the points we have to watch for.

We consider each circle separately.

a) The circle $|z - 1| = 1$ encloses the point $z_0 = 1$ where $g(z)$ is not analytic. We write $g(z)$ as

$$g(z) = \frac{z^2 + 1}{z^2 - 1} = \frac{z^2 + 1}{z + 1} \frac{1}{z - 1}$$

thus

$$f(z) = \frac{z^2 + 1}{z + 1}$$

Using

$$\oint_C \frac{f(z)}{z - z_0} dz = 2\pi i f(z_0)$$

we get

$$\oint_C \frac{z^2 + 1}{z^2 - 1} dz = 2\pi i f(1) = 2\pi i \left(\frac{z^2 + 1}{z + 1}\right)\bigg|_{z = 1} = 2\pi i$$

b) gives the same result as (a), since both closed path encloses the same complex number.

c)

$$g(z) = \frac{z^2 + 1}{z^2 - 1} = \frac{z^2 + 1}{z - 1} \frac{1}{z + 1}$$

$$f(z) = \frac{z^2 + 1}{z + 1}$$

Using

$$\oint_C \frac{f(z)}{z-z_0}dz = 2\pi i f(z_0)$$

we get

$$\oint_C \frac{z^2+1}{z^2-1}dz = 2\pi i f(-1) = 2\pi i \left(\frac{z^2+1}{z-1}\right)\bigg|_{z=-1} = -2\pi i$$

d) gives 0, since $g(z)$ is analytic inside d.

The integrals around each circle can be calculated using the MATLAB program MCode-4.14.

MCode-4.14

```
%A---------------------------------------------------
myFunc=@(z)((z.^2+1)./(z.^2-1));
g=@(theta)(1+cos(theta)+i*sin(theta));
gp=@(theta)(-sin(theta)+i*cos(theta));

result=integral(@(t) myFunc(g(t)).*gp(t),0, 2*pi)  % 6.2832i

2*pi*i;  % 6.2832i

%B---------------------------------------------------
myFunc=@(z)((z.^2+1)./(z.^2-1));
g=@(theta)(0.5+cos(theta)+i*sin(theta));
gp=@(theta)(-sin(theta)+i*cos(theta));

result=integral(@(t) myFunc(g(t)).*gp(t),0, 2*pi)  % 6.2832i

%C---------------------------------------------------
myFunc=@(z)((z.^2+1)./(z.^2-1));
g=@(theta)(-1+cos(theta)+i*(0.5+sin(theta)));
gp=@(theta)(-sin(theta)+i*cos(theta));

result=integral(@(t) myFunc(g(t)).*gp(t),0, 2*pi)  % - 6.2832i

%D---------------------------------------------------
myFunc=@(z)((z.^2+1)./(z.^2-1));
g=@(theta)(cos(theta)+i*(0.5+sin(theta)));
gp=@(theta)(-sin(theta)+i*cos(theta));

result=integral(@(t) myFunc(g(t)).*gp(t),0, 2*pi)  % 0
```

4.6.2 Multiply Connected Domains

If $f(z)$ is analytic on in the ring-shaped domain including its boundaries C_1 and C_2 as shown in Figure-4.16, and z_0 is an interior point of the domain, the value of the function at point z_0 can be evaluated as

$$f(z_0) = \frac{1}{2\pi i}\oint_{C_1} \frac{f(z)}{z-z_0} dz - \frac{1}{2\pi i}\oint_{C_2} \frac{f(z)}{z-z_0} dz$$

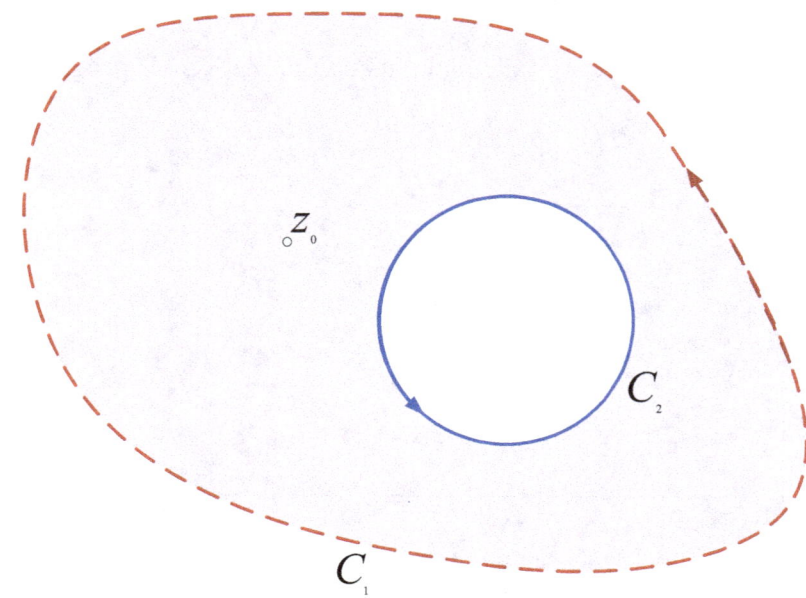

Figure-4.16

If the paths C_2 has opposite direction compared to C_1, then we have

$$f(z_0) = \frac{1}{2\pi i}\oint_{C_1} \frac{f(z)}{z-z_0} dz + \frac{1}{2\pi i}\oint_{C_2} \frac{f(z)}{z-z_0} dz$$

Example-4.15: Let

$$f(z) = z^2 \qquad g(z) = \frac{f(z)}{(z-1.5+j)} \qquad z_0 = 1.5 - j$$

Verify

$$f(z_0) = \frac{1}{2\pi i}\oint_{C_1} \frac{f(z)}{z-z_0} dz + \frac{1}{2\pi i}\oint_{C_2} \frac{f(z)}{z-z_0} dz$$

writing a MATLAB code for the complex region shown in Figure-4.17.

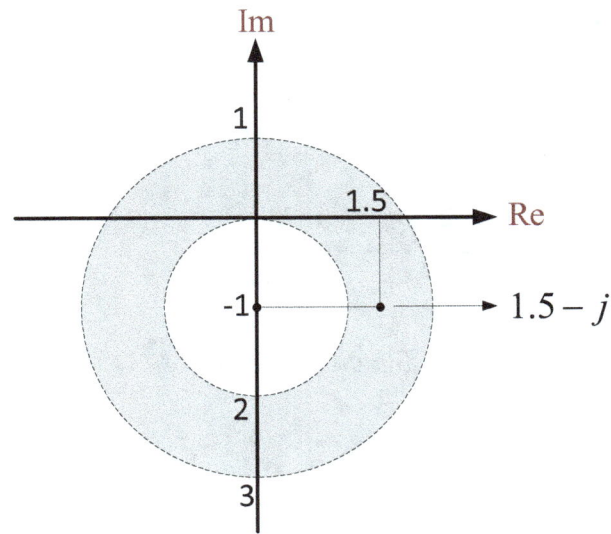

Figure-4.17

MCode-4.15

```
myFunc=@(z)(z.^2./(z-1.5+j) );
g=@(theta)(2*cos(theta)+i*(-1+2*sin(theta))); % C1
gp=@(theta)(-2*sin(theta)+i*2*cos(theta));
result1=integral(@(t) myFunc(g(t)).*gp(t),0, 2*pi); %

myFunc=@(z)(z.^2./(z-1.5+j) );
g=@(theta)(cos(theta)+i*(-1+sin(theta))); % C2
gp=@(theta)(-sin(theta)+i*cos(theta));
result2=integral(@(t) myFunc(g(t)).*gp(t),2*pi,0); %

result = (1./((2*pi)*j))*(result1+result2) %  1.2500 - 3.0000i
f_z0 = (1.5-j).^2 %  1.2500 - 3.0000i
```

4.7 Residue Integration

Let $f(z)$ be a complex function such that

$$f(z) = \frac{p(z)}{q(z)}$$

The roots of $p(z) = 0$ are called the zeros $f(z)$ and the roots of $q(z)$ are called the poles of $f(z)$.

The function $f(z)$ can be written as

$$f(z) = k\frac{(z-z_0)(z-z_1)\ldots(z-z_m)}{(z-p_0)(z-p_1)\ldots(z-p_n)}$$

which can be written after polynomial division as

$$f(z) = r(z) + \frac{b_0}{z-p_0} + \frac{b_1}{z-p_1} + \frac{b_2}{z-p_2} + \cdots + \frac{b_n}{z-p_n}$$

where the coefficients $b_0, b_1, \ldots, b_n$ are called the residues of $f(z)$

Note that the function $f(z)$ has no derivative at the poles.

The residues can be calculated as

$$b_i = \lim_{z \to p_i} (z-p_i)f(z)\big|_{z=p_i}$$

If $f(z)$ contains multiple poles than it can be written as

$$f(z) = r(z) + \sum_{i,k} \frac{b_i}{(z-p_i)^{k_j}}$$

The pole p_i has multiplicity of m, then the residue b_i at

$$\frac{b_i}{z-p_i}$$

is calculated as

$$b_i = \frac{1}{(m-1)!}\lim_{z \to p_i}\left\{\frac{d^{m-1}}{dz^{m-1}}\left[(z-p_i)^m f(z)\right]\right\}$$

Example-4.16:
$$f(z) = \frac{z+3}{z(z+2)^2(z+5)}$$
Find residues of $f(z)$

Solution-4.16: $f(z)$ can be written as

$$f(z) = \frac{r_1}{z} + \frac{r_2}{z+2} + \frac{r_3}{(z+2)^2} + \frac{r_4}{z+5}$$

r_1, r_3 and r_4 are founds as

$$r_1 = zf(z)|_{z=0} = \frac{z+3}{(z+2)^2(z+5)}\bigg|_{z=0} = \frac{3}{20}$$

$$r_3 = (z+2)^2 f(z)|_{z=-2} = \frac{z+3}{z(z+5)}\bigg|_{z=-2} = -\frac{1}{6}$$

$$r_4 = (z+5)f(z)|_{z=-5} = \frac{z+3}{z(z+2)^2}\bigg|_{z=-5} = \frac{2}{45}$$

$$r_2 = \left[\frac{d}{dz}(z+2)^2 f(z)\right]\bigg|_{z=-2}$$

$$= \left[\frac{d}{dz}\frac{z+3}{z(z+5)}\right]\bigg|_{z=-2}$$

$$= \frac{z(z+5) - (z+3)(2z+5)}{(z(z+5))^2}\bigg|_{z=-2} = -\frac{7}{36}$$

Using the calculated residues, we can write the function $f(z)$ as

$$f(z) = \frac{3}{20}\frac{1}{z} - \frac{7}{36}\frac{1}{z+2} - \frac{1}{6}\frac{1}{(z+2)^2} + \frac{2}{45}\frac{1}{z+5}$$

We can find the residues using MATLAB. For this purpose we first write the polynomial

$$f(z) = \frac{z+3}{z(z+2)^2(z+5)}$$

as

$$f(z) = \frac{z+3}{z^4 + 9z^3 + 24z^2 + 20z}$$

and using this form of $f(z)$ we can write a MATLAB program as in MCode-4.16 to find the residues.

MCode-4.16

```
b = [1 3];
a = [1 9 24 20 0];
[r,p,k] = residue(b,a)

x=[2/45 -7/36  -1/6 3/20]
```

125

```
% x =
%         0.0444   -0.1944   -0.1667    0.1500
```

When the program is run, we get the residues, poles and constant term as

r =	p =	k =
0.0444	-5.0000	[]
-0.1944	-2.0000	
-0.1667	-2.0000	
0.1500	0	

4.7.1 The Residue Theorem

Let $f(z)$ be an analytic function in a simple connected domain except for a number of poles as shown in Figure-4.18 and let C be a counterclockwise counter enclosing a number of poles, then, we have

$$\int_C f(z)dz = 2\pi j \sum_k b_k$$

where b_k is the residue of the pole p_k inside C.

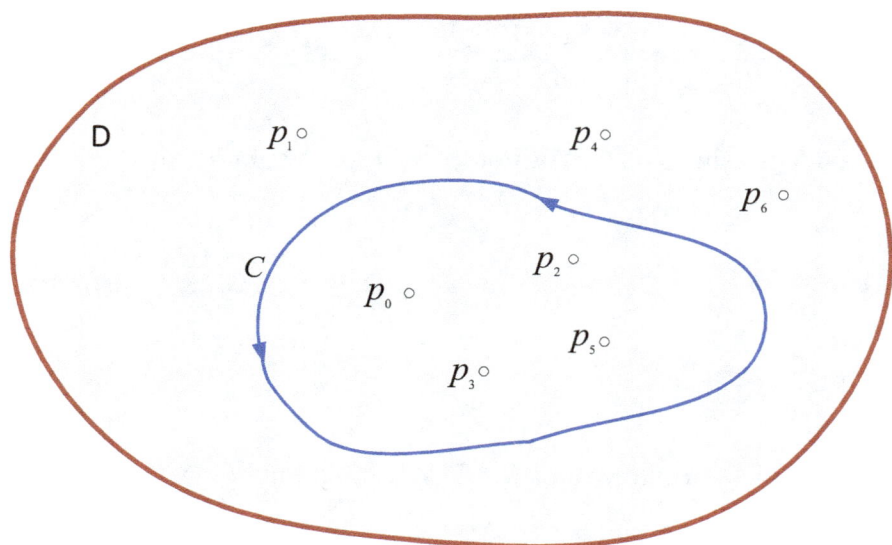

Figure-4.18

Example-4.17: Evaluate the contour integral

$$\oint_C \frac{dz}{z(1+z)}$$

If C is

a) the circle $|z| = 1/2$
b) the circle $|z| = 2$

Solution-4.17:

The poles, i.e., singularities of
$$f(z) = \frac{1}{z(1+z)}$$
are at $z = 0$ and $z = -1$

We can evaluate the residues as

Residue at $z = 0$ is $(z-0)f(z)|_{z=0} = 1$

Residue at $z = 1$ is $(z-(-1))f(z)|_{z=-1} = -1$

a) If C is $|z| = 1/2$ then it contains the pole at $z = 0$, but not the pole at $z = -1$. Hence, by the residue theorem

$$\oint_C \frac{dz}{z(1+z)} = 2\pi i \times (Residue\ at\ z = 0) = 2\pi i$$

b) If C is $|z| = 2$ then both poles are inside C. Hence, by the residue theorem

$$\oint_C \frac{dz}{z(1+z)} = 2\pi i \times (1-1) = 0$$

We can verify the obtained result using the MATLAB program MCode-4.17.

MCode-4.17

```
myFunc=@(z)(1./(z.*(z+1)));
g=@(theta)0.5*(cos(theta)+i*sin(theta));
gp=@(theta)0.5*(-sin(theta)+i*cos(theta));

result=integral(@(t) myFunc(g(t)).*gp(t),0, 2*pi) % 6.2832i

2*pi*i*r(2) % 6.2832i

%-----------------------------------------------
myFunc=@(z)(1./(z.*(z+1)));
```

```
g=@(theta)2*(cos(theta)+i*sin(theta));
gp=@(theta)2*(-sin(theta)+i*cos(theta));

result=integral(@(t) myFunc(g(t)).*gp(t),0, 2*pi)  % 0

2*pi*i*(r(1)+r(2))  % 0
```

4.8 Derivative and Limits of Complex functions

Open annulus in the complex plane can be expressed using the complex inequality

$$\rho_1 < |z - z_0| < \rho_2$$

where ρ_1, ρ_2 are the radiuses of the inner and outer circles and z_0 is the center of circles.

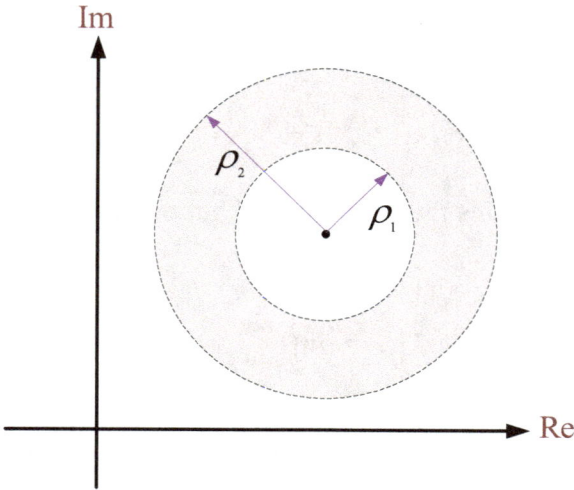

Figure-4.19

Closed annulus is expressed as

$$\rho_1 \leq |z - z_0| \leq \rho_2$$

We call circular disk $|z - z_0| < \rho$ as $\rho - neighborhood$ of z_0

In MCode-4.18, we draw the annulus $1 \leq |z| \leq 2$ in MATLAB.

MCode-4.18

```
% Number of complex numbers in a row of complex plane:
N = 50;

real_axis = linspace(-4, 4, N);
imag_axis = linspace(-4, 4, N);
[x, y] = meshgrid(real_axis, imag_axis);

% Generate a grid of complex numbers:
z = x + j * y;

plot(real(z), imag(z), 'k .')
xlabel('Real Axis');
ylabel('Imaginary Axis');
title('Complex Number Plane');
grid on;

c = (1<=abs(z)) & (abs(z)<=2) ;

hold on;
plot(real(z(c)), imag(z(c)), 'r *');
title('Complex plane set for 1<=|z|<=2');
```

When MCode-4.18 is run, we get the graph in Figure-4.20.

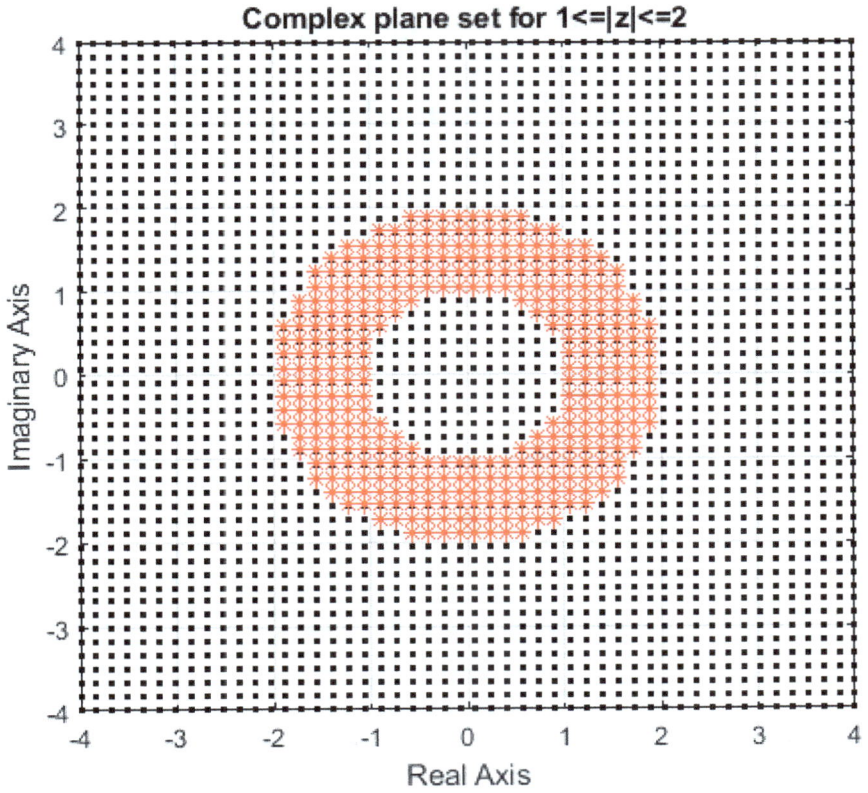

Figure-4.20

4.8.1 Half-Plane Definitions

Let $z = x + iy$, the open *upper* **half-plane** is defined as $y > 0$. The *lower half-plane* is defined by $y < 0$.

The *right half-plane* is defined by $x > 0$, and we define the *left half-plan* by $x < 0$

4.8.2 Complex Function

A complex function is defined as
$w = f(z)$

where z and w represents complex numbers.

4.8.3 Limit, Continuity

The limit of $f(z)$ is the complex number l and it is obtained as z approaches point z_0, and this is expressed as

$$\lim_{z \to z_0} f(z) = l$$

which implies that

$|f(z) - l| < \epsilon \qquad |z - z_0| < \delta$

This is graphically illustrated in Figure-4.21

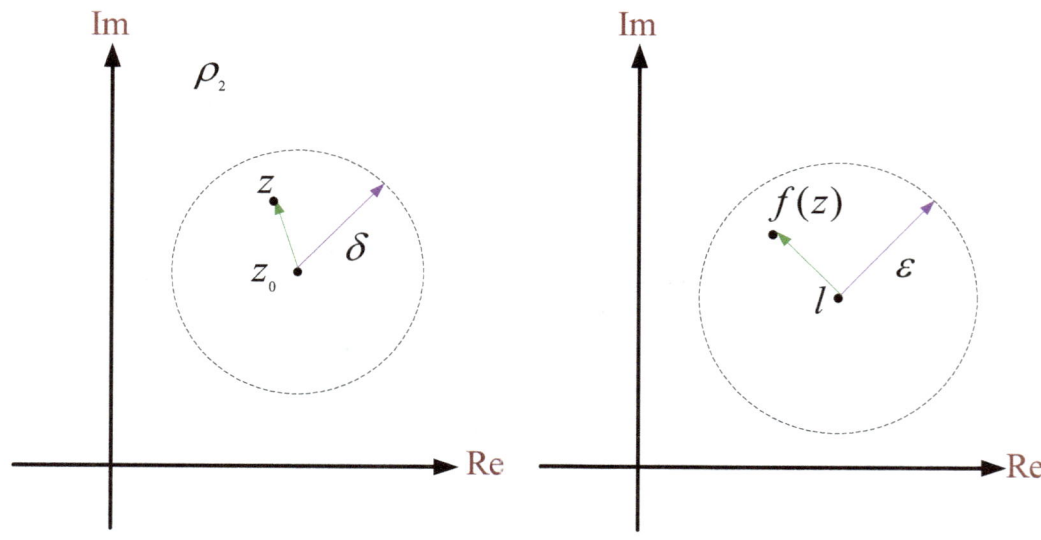

Figure-4.21

A function $f(z)$ is said to be **continuous** at $z = z_0$ if we have

$$\lim_{z \to z_0} f(z) = f(z_0)$$

$f(z)$ is said to be continuous in a domain if it is continuous at each point of this domain.

Derivative

The derivative of a complex function f at a point z_0, $f'(z_0)$, is defined by

$$f'(z_0) = \lim_{\Delta z \to 0} \frac{f(z_0 + \Delta z) - f(z_0)}{\Delta z}$$

where using $\Delta z = z - z_0$ we get

$$f'(z_0) = \lim_{z \to z_0} \frac{f(z) - f(z_0)}{z - z_0}$$

Example-4.18: Show that $f(z) = \bar{z}$ is not differentiable.

Solution-4.18: Let $z = x + iy$, $f(z) = x + iy$, $f(\Delta z) = \Delta x + i\Delta y$,

then $f(z) = \bar{z} = x - iy$ and using

$$f'(z_0) = \lim_{\Delta z \to 0} \frac{f(z_0 + \Delta z) - f(z_0)}{\Delta z}$$

we get

$$f'(z_0) = \lim_{\Delta z \to 0} \frac{\overline{(z_0 + \Delta z)} - \bar{z_0}}{\Delta z}$$

$$f'(z_0) = \lim_{\Delta z \to 0} \frac{\overline{\Delta z}}{\Delta z}$$

where using

$$\frac{\overline{\Delta z}}{\Delta z} = \frac{(\Delta x - i\Delta y)}{\Delta x + i\Delta y}$$

we get

$$f'(z_0) = \lim_{\Delta z \to 0} \frac{(\Delta x - i\Delta y)}{\Delta x + i\Delta y}$$

If $\Delta x = 0$, the result is -1 and

If $\Delta y = 0$, the result is 1

Thus, the limit does not exist.

4.9 Cauchy–Riemann Equations for Complex Numbers

Let $z = x + iy$, and the complex function $f(z)$ is defined as

$$f(z) = u(x,y) + i\, v(x,y)$$

If the function $f(z)$ is differentiable in a domain, then we have

$$\frac{\partial u(x,y)}{\partial x} = \frac{\partial v(x,y)}{\partial y} \qquad \frac{\partial u(x,y)}{\partial y} = -\frac{\partial v(x,y)}{\partial x}$$

and these equations are called Cauchy–Riemann equations, and the derivative of the function can be obtained using either

$$f'(z) = \frac{\partial u(x,y)}{\partial x} + i\frac{\partial v(x,y)}{\partial x} \rightarrow f'(z) = u_x + iv_x$$

or

$$f'(z) = -i\frac{\partial u(x,y)}{\partial y} + \frac{\partial v(x,y)}{\partial y} \rightarrow f'(z) = -iu_y + v_y$$

Example-4.19: The function

$$f(z) = z^2 = x^2 - y^2 + 2i\, xy$$

is analytic for all z, i.e., it is differentiable for all z, the Cauchy–Riemann equations satisfy, i.e.,

$$\frac{\partial u(x,y)}{\partial x} = \frac{\partial v(x,y)}{\partial y} \rightarrow \frac{\partial [x^2 - y^2]}{\partial x} = \frac{\partial [2xy]}{\partial y} \rightarrow 2x = 2x$$

$$\frac{\partial u(x,y)}{\partial y} = -\frac{\partial v(x,y)}{\partial x} \rightarrow \frac{\partial [x^2 - y^2]}{\partial y} = -\frac{\partial [2xy]}{\partial x} \rightarrow -2y = -2y$$

The derivative of the function is

$$f'(z) = 2z \rightarrow f'(z) = 2(x + iy) \rightarrow f'(z) = 2x + i2y$$

The derivative of the function can also be obtained either as

$$f'(z) = u_x + iv_x \to f'(z) = \frac{\partial[x^2 - y^2]}{\partial x} + i\frac{\partial[2xy]}{\partial x}$$

$$f'(z) = 2x + i2y$$

or as

$$f'(z) = -u_y + v_y \to f'(z) = -i\frac{\partial[x^2 - y^2]}{\partial y} + \frac{\partial[2xy]}{\partial y}$$

$$f'(z) = 2x + i2y$$

Bibliography

1) J.W. Brown, R. V. Churchill, Complex Variables, And Applications, McGraw-Hill Education, 2014, ISBN 978-0-07-338317-0

2) Glyn James et all, Advanced Modern Engineering Mathematics, Pearson Education, 2011, ISBN: 978-0-273-71923-6